Microplastic - An Unknown Danger

Preface to the Second Edition

Since the first edition of *Microplastic - An Unknown Danger* was published in 2024, the world's understanding of microplastics has deepened at an astonishing pace. What was once an emerging concern has now become a global priority for scientists, policymakers, and communities alike. In the short span of a year, groundbreaking research has revealed even more about how microplastics permeate our environment, our food, and even our bodies.

This **second edition** reflects the latest wave of scientific discoveries and societal awareness. I have expanded the book to include new studies that uncover microplastics in places we never imagined: in our blood, our organs, and even the placentas of newborns. Recent research has also linked microplastics to pressing health concerns, such as cardiovascular disease, preterm birth, and neurological impacts. We now know more about how everyday items-from non-stick cookware to synthetic clothing-contribute to this invisible threat, and about the potential for innovative solutions like biodegradable plastics to offer hope for a cleaner future.

Yet, the central questions remain: Is plastic a blessing or a curse? How do we balance its undeniable convenience with its hidden costs? And, most importantly, what can each of us do to confront this challenge?

This edition aims not only to inform, but to empower. By weaving together the latest research, real-world examples, and practical solutions, I hope to inspire readers to see microplastics not just as an abstract problem, but as a call to action for our health and our planet.

As we turn these pages together, let us continue to reflect, question, and strive for a world where the dangers of microplastics are no longer unknown, but understood-and ultimately, overcome.

Let us begin.

P.ABDUR RAHMAN

For questions or to learn more about microplastics, please feel free to contact me:
Email: pabrahman@yahoo.com
YouTube Channel:

https://www.youtube.com/@Microplastic-AnUknownDanger

Chapter 1: The Plastic Era and Its Impact

- Action versus Reaction (Page 7)
- Plastic Era (Page 8)

1.1 The Impact of Plastic on Various Industries (Page 11)

- Electrical Industry (11)
- Automobile Industry (15)
- Electronics Industry (16)
- Healthcare Industry (17)
- Packaging Industry (18)
- Construction Industry (19)
- Consumer Goods Industry (19)
- Agriculture Industry (23)

Chapter 2: Plastics – Manufacturing and Types

- Overview on Manufacturing and Types of Plastics (Page 30)
- Manufacturing (30)
- Types of Plastics (31)
- Thermoplastics (31)
- Polyethylene Terephthalate (PET or PETE) (31)
- High-Density Polyethylene (HDPE) (32)
- Low-Density Polyethylene (LDPE) (32)
- Polyvinyl Chloride (PVC) (32)
- Polypropylene (PP) (33)
- Polystyrene (PS) or Expanded Polystyrene (EPS) (33)
- Polycarbonate (PC) (34)
- Acrylonitrile Butadiene Styrene (ABS) (34)
- Polylactic Acid (PLA) (34)
- Nylon (Polyamide) (35)
- Thermosetting Plastics (35)
- Epoxy Resins (35)
- Polyester Resins (36)

- Polyurethane Resins (36)
- Silicone Resins (36)
- Melamine Formaldehyde (37)
- Urea Formaldehyde (37)
- Diallyl Phthalate (38)
- Vinyl Esters (38)

Chapter 3: Microplastics – Formation, Detection, and Impacts

- Microplastic (Page 39
- Detection of plastic debris (39)
- Breakdown Processes Leading to Microplastics (40)
- Rate of Disintegration of Plastic (42)
- The Factors that Determine the Disintegration of Plastic (43)
- 3.1 Recent Scientific Findings on Microplastic Pollution and Its Impact
- Microplastics in Our Rain: An Unseen Storm (45)
- Oceanic and Marine Ecosystems (48)
- The Threat of Microplastics to Coral (50)
- Soil and Agricultural Impacts (51)
- Microplastics and Microbes: A Dangerous Alliance (57)
- The Silent Invasion: Microplastics in Human Organs (58)
- Microplastics in IV fluids (63)
- Microplastics in tea bag (65)
- Invisible Risks:Microplastic Release from Non-Stick Cookware(66)
- 3.2 Microbeads (68)

Chapter 4: Global Perspectives and Legislative Efforts

- Global Organisations' Report on Microplastic (Page 73)
- World Health Organisation Concerns Over Drinking Water (73)
- European Commission's Press Release on Restricting Microplastics (74)
- World Economic Forum's Warning on Microplastic (75)

- United Nations Environment Programme (UNEP) Report on Microbeads (76)
- Food and Agriculture Organisation of the United Nations's Report (77)
- Pathways of Microplastics Contamination (81)

Chapter 5: Plastics in Emerging Industries

- Plastic in Emerging Industries (Page 84)
- Artificial Intelligence (AI) (84)
- Cyber security (84)
- Electric Vehicles (EVs) (85)
- Renewable Energy (86)
- Blockchain and Cryptocurrency (86)
- Virtual and Augmented Reality (VR/AR) (87)
- 3D Printing (87)
- Biotechnology (88)
- Nanotechnology (88)
- Space Exploration (89)
- Telecommunication (89)

Chapter 6: Global Legislative Efforts on Reducing Plastic Pollution

- Global Legislative Efforts on Reducing Plastic Pollution (Page 92)
- United States of America (92)
- United Kingdom (93)
- European Union (EU) (93)
- Canada (94)
- Brazil (94
- China (95)
- Gulf Countries (UAE, Saudi Arabia, Qatar, etc.) (95)
- Australia (96)
- Japan (96)
- Rwanda (97)
- New Zealand (97)
- Singapore (97)

- Thailand (98)
- India (98)

Chapter 7: The Shift to Biodegradable Materials

- The Shift to Biodegradable Materials Across Industries (Page 100)
- Packaging Industry (100)
- Textiles and Fashion Industry (101)
- Consumer Goods (102)
- Construction and Building Materials (103)
- Agriculture (104)
- Healthcare (105)
- Fishing Industry (106)

Chapter 8: The Future – Challenges and Solutions

- Earth Exists Because of Extinction (Page 108)
- Tips to Reduce Microplastic Exposure (Page 109)
- Our Responsibilities (Page 116)
- Hopeful Advances in Biodegradable Plastics (Page 119)

Chapter 1
The Plastic Era and Its Impact

Action versus Reaction

"To every action, there is an equal and opposite reaction."

This is Newton's third law, a principle that applies not only to physical forces but also to the broader interaction between humans and nature. Just as every force in the physical world is met with an equal and opposite force, every action we take on the environment elicits a response from nature. Whatever we do to nature, it reacts with consequences of equal measure.

Every invention of humanity has its pros and cons in relation to nature. This can be seen when we examine history. Human inventions have shaped civilization, changing how we live, work, and interact.

In the early days, wheels for animal carts were made from wood, a natural resource, causing little environmental harm. As long as humans lived in harmony with nature, the reactions from nature remained subtle and often went unnoticed.

However, during the Industrial Revolution, the invention of machinery revolutionized production processes. Mass production became more efficient, reducing labour and manufacturing costs while increasing the availability of goods. While this progress benefited humanity, it also initiated a powerful reaction from nature.

In accordance with Newton's third law, the unchecked exploitation of natural resources has led to environmental degradation. Industrialization resulted in pollution, habitat destruction, and resource depletion.

As humans pushed nature beyond its limits, nature responded in turn, triggering severe consequences such as climate change, global warming, melting glaciers, rising sea levels, and the emergence of new pathogens.

Newton's law reminds us that no action exists in isolation. Just as objects in motion produce equal and opposite reactions, our actions toward nature bring about inevitable repercussions. The more we disrupt the environment, the stronger nature's response.

In that sense, plastic, an invention of humanity, was considered a valuable boon when it was first discovered. Now, plastic has become so deeply embedded in the products we use daily that we cannot even imagine a world without it.

Plastic Era

The invention of the first synthetic plastic, Bakelite, by Leo Baekeland in 1907 marked a pivotal moment in materials science. Mass production of plastics gained momentum during World War II, driven by their lightweight and durable properties, which made them ideal for military applications. For example, nylon was used in parachutes and ropes, while plexiglass replaced glass in aircraft windows due to its shatter-resistant nature.

In the post-war period, consumer demand for affordable and durable products surged. Plastics, being both versatile and cost-effective, quickly became the material of choice for manufacturers. The Plastic Era saw traditional materials like glass, metal, and wood being replaced by plastics in household goods. Today, global plastic production exceeds 400 million metric tons annually.

One of the key reasons for the widespread adoption of plastics is their unmatched versatility. Unlike metals or glass, plastics can be moulded into almost any shape and tailored to meet specific needs, such as flexibility, strength, or transparency. This adaptability makes plastics ideal for a wide range of applications—from lightweight packaging that preserves food freshness to intricate components used in the automotive and aerospace industries.

In addition, plastics are generally lighter than metals or glass, which helps reduce transportation costs and fuel consumption. Their resistance to corrosion and chemicals increases their longevity, making them appealing for both consumer and industrial use. Furthermore, plastic production is often less energy-intensive compared to traditional materials, contributing to their growing popularity.

World welcomed Plastic

Bakelite was positioned as an alternative to natural materials like ivory, wood, and tortoiseshell, which were becoming scarce due to overexploitation. This aspect was particularly appealing during a time of increasing awareness about resource conservation. During its early development, the world enthusiastically embraced plastic. Chemists were excited by the fact that plastics could be engineered from simple petrochemicals, opening the door to creating new materials with specific properties. The ability to design materials at the molecular level fascinated scientists.

By the mid-20th century, particularly in the 1950s and 1960s, other forms of synthetic plastics became cornerstones of industrial growth. Plastics were seen as modern and innovative, driving the production of consumer goods, packaging, and construction materials.

Plastics were marketed as "***miracle materials***" that could improve daily life. They were lightweight, inexpensive, durable, and easy to manufacture, making them highly appealing to consumers. Plastic containers, packaging, and household items quickly became widespread. People embraced plastics for their ability to create new, affordable products that were previously out of reach. Plastics also reduced dependency on finite natural resources, making them seem like a forward-thinking solution.

The phrase *"**the material of a thousand uses**"* was often associated with Bakelite, the first synthetic plastic, invented by Leo Baekeland in 1907. This slogan highlighted Bakelite's versatility across various industries, from household goods to electrical insulators and automotive parts.

Similarly, the slogan *"**Better Living Through Chemistry**,"* popularized by DuPont in the 1930s and 1940s, emphasized the benefits of chemical innovations, including how plastics and other chemicals improved everyday life.

Bakelite was first used in manufacturing electrical components due to its heat resistance and electrical insulating capabilities. This led to its widespread adoption in various electrical applications. The use of

Bakelite allowed for the production of more reliable and durable electrical instruments, enhancing the safety and efficiency of electrical systems.

This initial application in the electrical industry opened the door for Bakelite's use in other sectors, such as automotive, consumer goods, and household items, establishing it as a crucial material in the development of modern technology.

1.1. The Impact of Plastic on Various Industries

The invention of plastic revolutionized various industries, leading to unprecedented advancements in technology, design, and functionality. Introduced in the early 20th century, plastic quickly became a ubiquitous material, transforming everything from consumer goods to healthcare, and reshaping manufacturing processes. Its versatility, lightweight nature, durability, and cost-effectiveness have made it an essential component across diverse sectors.

1. Electrical Industry

Before the invention of plastic and other synthetic insulation materials, electrical wire connections were insulated using various natural materials. These materials provided the necessary protection to prevent electrical short circuits, overheating, and potential hazards. Some common materials used for insulation in the early days of electrical wiring included

i) Rubber

Natural rubber was one of the earliest and most effective insulators used in electrical wiring. It was flexible, had good insulating properties, and could protect wires from moisture. Rubber insulation became popular in the late 19th century and remained widely used.

Limitations

- **Degradation**: Natural rubber can deteriorate over time due to exposure to heat, sunlight, and ozone, leading to brittleness and loss of insulating properties.

- **Moisture Absorption**: Rubber can absorb moisture, which may compromise its insulating ability and increase the risk of electrical leakage.
- **Flammability**: Rubber is relatively flammable, posing a risk in high-temperature environments.

ii) Cloth and Silk

Before rubber and plastic, woven cloth, particularly cotton or silk, was used to wrap wires. These materials were coated with wax or varnish to enhance their insulating properties and protect the wires from external elements. Cloth insulation was common in low-voltage applications, but it wasn't as durable as rubber or later synthetic materials.

Limitations:

- **Durability**: Cloth and silk are less durable than rubber or synthetic materials, making them susceptible to wear and tear over time.
- **Flammability**: These materials are flammable, posing a fire hazard, especially in higher voltage applications.
- **Moisture Vulnerability**: While coatings provided some protection, cloth could still absorb moisture, leading to degradation and reduced insulating effectiveness.

iii) Asbestos

Description: Asbestos, a naturally occurring mineral, was used in some cases to insulate electrical wires because of its heat resistance and non-conductive properties. However, asbestos is now known to be hazardous to human health, and its use has been discontinued.

Limitations:

- **Health Hazards**: Asbestos is linked to serious health issues, including lung cancer and asbestosis, leading to strict regulations and a decline in its use.
- **Brittleness**: Asbestos insulation can become brittle over time, reducing its effectiveness and increasing the risk of exposure to harmful fibers.

- **Difficult Removal**: Removing asbestos insulation poses significant health risks and requires specialized handling.

iv) Gutta-Percha

Description: Gutta-percha, a natural latex derived from trees, was used in the mid-19th century as an insulator for telegraph wires and underwater cables. It was a durable, flexible material that provided effective insulation, particularly for long-distance applications like transatlantic cables.

Limitations

- **Environmental Sensitivity**: Gutta-percha can become soft and lose its insulating properties when exposed to high temperatures or certain chemicals.
- **Limited Applications**: While effective for some applications, it was not suitable for all types of wiring, particularly in environments with extreme temperatures.
- **Cost and Availability**: Gutta-percha can be more expensive and less readily available than other insulating materials.

v) Wood and Ceramic

Description: In some cases, wooden and ceramic insulators were used, particularly in early overhead power lines and telegraph systems. Wooden poles supported the wires, and ceramic insulators (like those seen in old telephone poles) were used to prevent the electricity from leaking or grounding through the supporting structures.

Limitations

- **Moisture Absorption**: Wood can absorb moisture, leading to decay and reduced structural integrity over time.
- **Brittleness of Ceramics**: Ceramic insulators can be brittle and prone to breaking under stress or impact.

- **Limited Voltage Ratings**: Wooden and ceramic insulators may not perform well at higher voltage levels, limiting their use in modern applications.

vi) Oil-Soaked Paper

For early power cables and transformers, oil-soaked paper was used as an insulating material. The oil acted as a dielectric to reduce the risk of electrical discharge, and the paper served as the structural medium to separate and insulate conductors. This was commonly seen in underground and submarine cables before synthetic alternatives were available.

Limitations

- **Degradation Over Time**: The paper can deteriorate with age, especially if exposed to moisture, leading to loss of insulation and potential short circuits.
- **Oil Leakage**: The oil can leak out, compromising the dielectric properties and requiring maintenance or replacement.
- **Environmental Concerns**: Disposal of oil-soaked materials poses environmental challenges due to potential contamination.

vii) Varnish and Resin Coatings

Wires were often coated in varnish or resins to provide a basic layer of insulation. These coatings were applied by dipping or brushing onto the wires to protect them from moisture and reduce electrical leakage. Varnish insulation was commonly used for winding in motors and transformers.

Limitations

- **Limited Insulating Properties**: While varnishes and resins provide some insulation, they may not be sufficient for high-voltage applications.
- **Wear Over Time**: Varnish can wear off or degrade over time, especially when exposed to heat or humidity, leading to reduced effectiveness.

- **Application Challenges**: Achieving a uniform coating can be difficult, and improper application can lead to weak points in insulation.

Each of these early insulation materials played a crucial role in the development of electrical wiring and devices. However, their limitations highlighted the need for more durable, reliable, and safer materials, ultimately paving the way for the widespread adoption of synthetic plastics in electrical and electronic applications.

Safety and Security

Even though plastics have fire resistance, chemical resistance, durability, and versatility, their insulation properties are the most welcomed characteristic. This property ensures safety and security when using electrical devices.

Electrical Parts of Bakelite

2. Automobile Industry

In the automobile sector, plastic plays a crucial role in enhancing fuel efficiency and vehicle safety. By replacing traditional materials like steel and aluminium, plastics help reduce the overall weight of vehicles, allowing for lower energy consumption and reduced emissions.

Plastics are moulded into complex shapes, enabling innovative designs that improve aerodynamics and aesthetics. Furthermore, plastic components are resistant to rust, ensuring durability in various environmental conditions.

From bumpers to dashboards, plastics contribute to both the longevity and safety of vehicles, showcasing their significant impact on modern automotive engineering.

Old Model cars

New Model Car

2. Electronics Industry

ENIAC

Old Model Computer

New model Computer

The electronics industry has greatly benefited from the advent of plastic. The first computers were primarily developed using metals for most of their structural components, while plastics were used in certain parts, particularly for insulation and some housing elements. Early computers like ENIAC (1945) and Colossus (1943) featured metal frames and components, such as vacuum tubes, relays, and metal wiring for circuits.

However, plastics like Bakelite were already in use for electrical insulation and non-conductive components, including casings, switch covers, and some wiring insulation. Plastics helped improve safety and reliability by insulating electronic components and preventing short circuits.

Because of their insulating properties and flexibility, plastics are perfect for many electronic components, helping to prevent short circuits and electrical hazards. Plastics can be moulded into complex shapes, encouraging innovation and the design of comfortable and attractive products. Without plastics, consumer electronics would be much heavier, more fragile, and less user-friendly.

4. Healthcare Industry

In healthcare, plastic has made remarkable contributions to patient care and safety. Disposable plastic items, such as syringes, gloves, and IV bags, minimize the risk of infection and ensure a higher standard of hygiene.

Plastics have transformed the healthcare industry in numerous ways, significantly improving patient care, safety, and efficiency. Here are some key areas where plastics have made a revolutionary impact:

Key Applications of Plastics in Medical Devices

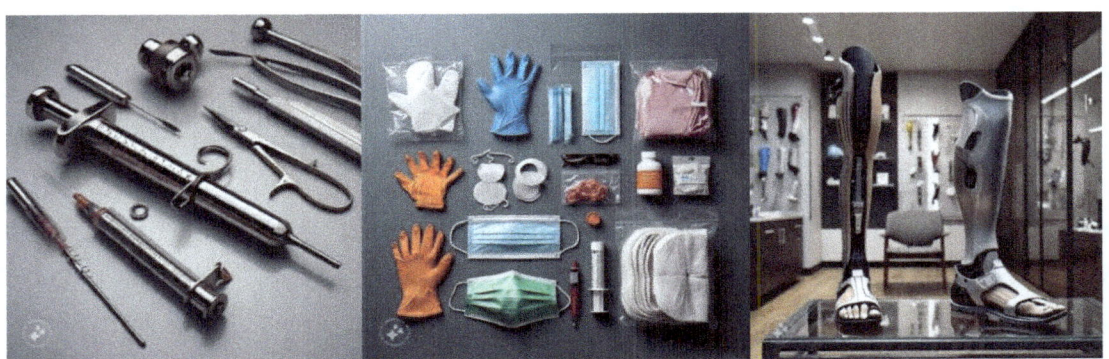

Metal medical equipment   disposable items   Prosthetics

i) **Syringes and IV Bags:** One of the most ubiquitous applications of plastics in healthcare is in the production of syringes and intravenous (IV) bags. Disposable plastic syringes ensure sterility and reduce the risk of infection, while IV bags made from flexible plastics facilitate the safe delivery of fluids and medications directly into a patient's bloodstream. The lightweight nature of these materials also makes them easy to handle and transport.

ii) **Diagnostic Devices:** Many diagnostic tools, such as blood glucose meters and blood pressure cuffs, incorporate plastic components. These devices benefit from the lightweight and durable characteristics of plastics, allowing for accurate and timely patient testing. The ease of sterilization further enhances their utility in clinical settings.

iii) **Prosthetics:** (*artificial limbs, such as hands or legs*) **and Orthotics** (*devices that support or correct existing limbs*)**:** The customization potential of plastics has transformed the field of prosthetics and orthotics. Lightweight, durable plastics are used to create individualized prosthetic limbs and orthotic devices, significantly improving mobility and quality of life for patients. Advances in material science continue to enhance the performance and comfort of these devices.

iv) **Infusion Pumps and Dialysis Equipment:** Infusion pumps, which deliver medications and nutrients through IV lines, and dialysis machines, which treat kidney failure, utilize plastic components extensively. These devices require materials that can withstand various conditions while ensuring patient safety and efficacy in treatment.

5. Packaging Industry

The packaging industry has undergone a major transformation due to plastic. Its lightweight and protective properties make it an ideal choice for packaging food, beverages, and consumer goods, safeguarding them from damage, contamination, and spoilage, which extends their shelf life.

The versatility of plastics allows for various forms, including flexible films, rigid containers, and protective wraps. This adaptability has not only improved supply chain efficiency but has also contributed to the growth of e-commerce by ensuring the safe delivery of products.

Old packing materials

New packing materials

6. Construction Industry

Pre-plastic House model

Plastic integrated Modern House

Plastic has been a game-changer in the construction industry, where its use in pipes, insulation, windows, and flooring has led to more efficient and durable building practices. Plastic materials are used for pipes and fittings, offering resistance to corrosion, chemicals, and water damage. This makes them ideal for plumbing, drainage, and electrical installations. Insulation materials made from plastic help improve energy efficiency in buildings, reducing heating and cooling costs.

Additionally, plastic is used in roofing, wall panels, and flooring due to its durability, moisture resistance, and ease of maintenance. The lightweight nature of plastic also reduces transportation costs and simplifies installation processes.

7. Consumer Goods Industry

Traditional House hold items

The consumer goods industry has greatly benefited from plastic's adaptability and affordability. Plastics are used in everyday items like toys, household appliances, furniture, and kitchenware. Their lightweight and durable nature makes them ideal for products that need to be inexpensive, easy to handle, and long-lasting.

Modern House hold items

The ability to produce plastics in various colours and finishes allows manufacturers to create appealing, customizable products that meet consumer preferences. This flexibility in design has fostered creativity and innovation in product development, significantly enhancing consumer choice. For example, plastic has replaced glass containers with plastic storage containers and bottles, as well as wooden utensils with plastic kitchenware.

From toothbrushes and storage containers to televisions and washing machines, plastic's presence in consumer goods has enabled companies to meet the growing demand for functional, attractive, and affordable products.

Before the invention of plastic packaging, snacks like biscuits and other food items were packed and sold using more traditional materials such as paper, cloth, and natural fibres. Here's a breakdown of how different consumer items, especially food, were packaged and sold in the olden days:

Biscuits and Snacks:

Paper Packaging: Biscuits were often wrapped in waxed paper or parchment paper to keep them fresh. In some cases, they were sold loose by weight, with shopkeepers packing them in paper bags for customers.

Tin Containers: Biscuits and other dry snacks were also sold in reusable tin containers or metal boxes, which customers could bring to refill, reducing the need for single-use packaging.

Glass Jars: Shopkeepers often kept biscuits and snacks in large glass jars with tight lids. Customers could ask for specific quantities, which were then packed in paper bags.

Grains, Flour, and Rice:

These were generally sold in cloth sacks or jute bags. The sacks were often reusable, and customers would bring their own containers to the market to fill up as needed.

Dairy Products:

Milk: Before plastic bottles, milk was delivered in glass bottles, which were sterilized and reused. In rural areas, people brought their own containers to buy milk.

Butter and Cheese: These were wrapped in waxed paper, muslin cloth, or stored in earthenware pots.

Fruits and Vegetables:

These were sold loose and weighed in baskets or directly handed to customers in cloth or paper bags. There was little to no packaging involved, and buyers often used baskets to carry them home.

Bread:

Bakers wrapped bread in paper or cloth. In some cases, bread was sold directly without wrapping, and customers used their own cloth bags to carry it.

Meat and Fish:

Butchers and fishmongers wrapped meat in thick brown paper or greaseproof paper. In some cases, customers brought their own containers

Sweets and Confectionery:

Sweets were usually stored in glass jars or tin containers. When sold, they were weighed and placed in small paper bags.

Tea, Coffee, and Spices:

These items were typically stored in metal tins or wooden boxes and sold loose by weight, wrapped in paper.

Challenges in Olden Days

- **Shorter Shelf Life**: Without plastic, food items, especially snacks like biscuits, would go stale faster. Waxed paper and metal tins offered some protection, but keeping products fresh for long was a challenge.
- **Limited Portability**: Cloth and paper could not protect against moisture as well as modern plastic, so items could get damaged during transport or in humid conditions.
- **Environmental Impact**: Though biodegradable, materials like paper, jute, and tin required more effort to produce and were sometimes less efficient in terms of storage and protection.

Later, plastic bags and containers emerged to address these difficulties, significantly increasing the shelf life of food items and protecting them from damage during transport.

8. Agriculture Industry

Everyone in this world depends on food. Energy is required for us to perform our routine activities, and food provides the necessary energy to keep us engaged and active.

As the population increases, so does the demand for food. Plastic has played a significant role in addressing this challenge. From improving crop yields to reducing labour costs, plastic has been vital in making farming more efficient and productive.

Here, we will explore some of the best applications of plastic in agricultural activities.

Plastic Mulch and Weed Control

One of the most impactful innovations in agriculture is the use of plastic mulch. This thin plastic sheet is spread over soil beds to suppress weed growth, conserve soil moisture, and regulate temperature. Before the advent of plastic, farmers relied on labour-intensive weed control methods that were often less effective.

With plastic mulch, weeds are kept at bay without the need for chemical herbicides, and the soil remains moist, reducing the need for frequent irrigation. The controlled temperature created by plastic mulch accelerates plant growth, leading to higher crop yields and more efficient farming practices.

Greenhouses and Controlled Environment Agriculture

The development of plastic films for greenhouses has transformed agriculture by allowing farmers to grow crops in controlled environments. These plastic greenhouses provide optimal growing conditions by shielding plants from extreme weather, pests, and diseases. The plastics used in greenhouse construction are lightweight, affordable, and easy to install, making this technology accessible to a wide range of farmers.

With greenhouses, agriculture is no longer limited to specific growing seasons or climates. Farmers can now cultivate fruits, vegetables, and flowers year-round, regardless of external weather conditions. This extension of the growing season significantly impacts food supply chains and global food security.

Limitations: While greenhouses are ideal for smaller plants, fruits, and vegetables, they are not suited for larger trees like coconut palms, which require significant space and depth for their root systems.

- **Vegetables**: Tomatoes, cucumbers, peppers, lettuce, and herbs thrive in greenhouse environments.
- **Fruits**: Strawberries, melons, and citrus fruits can be cultivated effectively.
- **Flowers**: Cut flowers, potted plants, and ornamental varieties are commonly grown in greenhouses.

Irrigation Systems

Plastic has also played a critical role in modern irrigation systems. Drip irrigation, in particular, utilizes plastic tubing and emitters to deliver water directly to plant roots, minimizing water waste. This technology is especially beneficial in arid regions where water is scarce.

The use of plastic in irrigation reduces evaporation, soil erosion, and nutrient runoff, which enhances soil fertility and ensures that crops receive the appropriate amount of water. These systems also reduce labour costs and increase productivity by automating the watering process.

Seed Coating:

Seed coating is an agricultural technique in which a protective layer is applied to seeds before sowing. This coating typically consists of a mixture of materials, including polymers, fertilizers, pesticides, and sometimes beneficial microorganisms.

Seed coating enhances seed performance and ensures better germination and growth under various environmental conditions. All of the theses implementations have resulted in higher crop yields. The quantity of agricultural products produced has significantly increased.

Plastic has significantly contributed to agriculture in various ways beyond plastic mulch, greenhouses, drip irrigation and seed coating. Here are some additional applications:

Protective Row Covers:

Lightweight plastic covers protect plants from pests, frost, and harsh weather conditions. They create a microclimate that can enhance growth and yield.

Plant Pots and Containers:

Plastic pots and trays are lightweight, durable, and often reusable, making them ideal for nurseries and transplanting. They help in efficient root development and ease of handling.

Water Management:

Plastic liners and membranes are used in ponds and reservoirs to reduce evaporation and prevent water contamination, ensuring a reliable water supply for irrigation.

Harvesting and Packaging:

Plastic containers and bags are commonly used for harvesting and transporting produce. They help maintain freshness and reduce spoilage during transport.

Pest Control:

Plastic barriers, such as insect nets, are effective for preventing pest entry while allowing sunlight and rain to reach the crops.

Aquaculture:

Plastic materials are used in aquaculture systems for creating tanks, nets, and other equipment, promoting fish farming practices.

Chapter 2
Plastics – Manufacturing and Types

Overview on Manufacturing and Types of Plastics
Manufacturing:

i) **Raw Materials:** Plastics are manufactured using petrochemicals, which are products of crude oil or natural gas. Natural gas is a source of ethane and propane, which can be converted into ethylene and propylene.

ii) **Polymerization**: The extracted hydrocarbons (monomers) undergo polymerization or poly condensation processes. In polymerization, monomers chemically bond to form long polymer chains, creating plastics such as polyethylene, polypropylene, and PVC.

Here's a breakdown of the key hydrocarbons used for producing plastics of different kinds.

Hydrocarbon	Type of plastic
Ethylene	Used to produce polyethylene(PE)
Propylene	The base for polypropylene (PP)
Styrene	Forms polystyrene (PS)
Vinyl Chloride	Used to produce polyvinyl chloride (PVC)
Terephthalic Acid / Ethylene Glycol	Essential for producing polyethylene terephthalate (PET)
Butadiene	Used to produce acrylonitrile butadiene styrene (ABS)
Phenol	Combined with formaldehyde to produce phenolic resins
Formaldehyde (CH_2O)	Used to produce melamine and urea-formaldehyde resins
Epichlorohydrin	A key precursor for epoxy resins
Isocyanates - Particularly Toluene di isocyanate (TDI) and Methylene di phenyl di isocyanate (MDI)	Used to make polyurethanes.
Maleic Anhydride	An important compound in the production of unsaturated polyester resins.

iii) Apart from these raw materials, various additives are included to enhance the properties of plastics:

- **Plasticizers**: Improve flexibility (e.g., phthalates).
- **Stabilizers**: Protect against UV light and heat degradation.
- **Fillers**: Enhance strength or reduce costs (e.g., talc, calcium carbonate).
- **Colourants**: Provide desired colours to the final product.

iv) Moulding/Extrusion: The plastic is melted and shaped through processes such as injection moulding, extrusion, or blow moulding, depending on the desired product (e.g., bottles, films, or parts).

v) Cooling and Finishing: The formed plastic is cooled, solidified, and further processed through trimming, polishing, or assembly into final products.

Types of Plastics:

The two main categories of plastics are thermoplastics and thermosetting plastics (thermosets), with the key difference being that a thermoset cannot be re-melted, while a thermoplastic can.

Thermoplastics include, among others, polyethylene (PE), polypropylene (PP), polyvinyl chloride (PVC), polyethylene terephthalate (PET), polystyrene (PS), polycarbonate (PC), and polyamide (PA). The main types of thermosets are polyurethane (PU), epoxy resins, vinyl esters, and silicones.

Thermoplastics:
Polyethylene Terephthalate (PET or PETE)

Food containers

Beverage bottle

- **Properties**: Lightweight, strong, and resistant to moisture.
- **Uses**: Beverage bottles, food containers, synthetic fibres (e.g., polyester).

2. High-Density Polyethylene (HDPE)

- **Properties**: Durable, strong, and resistant to chemicals.
- **Uses**: Milk jugs, detergent bottles, water pipes

Detergent bottles

Milk jug

3. Low-Density Polyethylene (LDPE)

- **Properties**: Flexible, lightweight, and transparent.
- **Uses**: Plastic bags, food wraps, squeezable bottles.

Squeezable bottles

Plastic bags

4. Polyvinyl Chloride (PVC)

- **Properties**: Rigid or flexible, resistant to weathering, and non-flammable.
- **Uses**: Plumbing pipes, window frames, flooring, medical equipment.

Plumbing pipes

Window frames

5. Polypropylene (PP)

- **Properties**: Heat-resistant, durable, and flexible.
- **Uses**: Food containers, automotive parts, bottle caps, packaging tapes.

Packaging tapes

Bumpers

6. Polystyrene (PS) or Expanded Polystyrene (EPS)

- **Properties**: Lightweight, rigid, or foam-like (expanded form).
- **Uses**: Disposable coffee cups, food trays, packing materials, insulation.

Disposable coffee cups & tray   Packing materials

7. Polycarbonate (PC)

- **Properties**: Transparent, strong, and impact-resistant.
- **Uses**: Eyeglass lenses, water bottles, electronic devices, Compact Disc / (CD) / Digital Versatile Disc (DVD).

Eyeglass lenses

CD / DVD

8. Acrylonitrile Butadiene Styrene (ABS)

- **Properties**: Tough, impact-resistant, and durable.
- **Uses**: Toys (e.g., LEGO bricks), automotive parts, electronics housings.

LEGO Bricks

Automotive part

9. Polylactic Acid (PLA)

- **Properties**: Biodegradable, made from renewable resources like corn starch
- **Uses**: Food packaging, disposable tableware, cutlery, biodegradable bags. 3D Printing.

Disposable tableware

Cutlery

10. Nylon (Polyamide)

- **Properties**: Strong, elastic, and resistant to wear and chemicals.
- **Uses**: Clothing, fishing nets, ropes, and automotive components.

Nylon rope

Fishing net

Thermosetting Plastics:

1. Epoxy Resins

Properties: High strength, excellent chemical resistance and good thermal stability.
Uses: Adhesives, coatings, and composite materials.

Adhesive

2. Polyester Resins

- **Properties**: Good chemical resistance and dimensional stability; often used in composite materials.
- **Uses**: Fibreglass, automotive parts, and marine applications

Fibreglass

3. Polyurethane Resins

- **Properties**: Good flexibility, toughness, and chemical resistance, depending on formulation.
- **Uses**: Rigid foams, coatings, and adhesives.

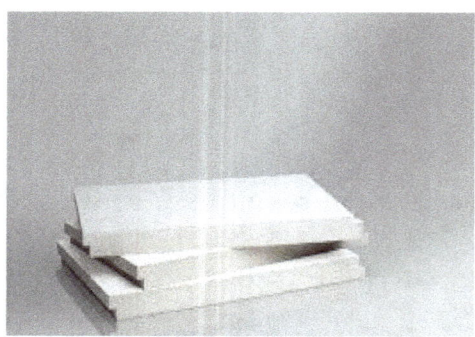

Rigid Foam

Adhesive

4. Silicone Resins

- **Properties**: Excellent temperature resistance and flexibility.
- **Uses**: Sealants, coatings, and high-temperature applications.

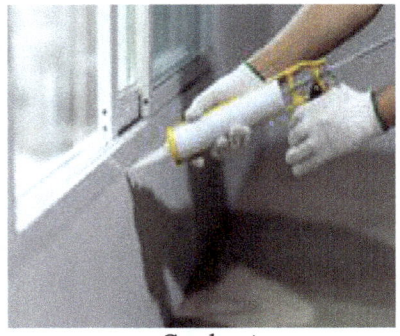
Sealant

Coating

5. Melamine Formaldehyde

- **Properties**: Hard, scratch-resistant surface with good heat resistance.
- **Uses**: Laminates, dinnerware, and adhesives.

Dinnerware

Laminate

6. Urea Formaldehyde

- **Properties**: Strong and rigid, with good electrical insulating properties.
- **Uses**: Particle board, adhesives, and moulded products.

Particle board

Adhesive

7. Diallyl Phthalate

- **Properties**: Good heat and chemical resistance.
- **Uses**: Electrical applications and automotive parts.

Electrical parts

Automotive parts

8. Vinyl Esters

- **Properties :** Exhibits excellent tensile strength and impact resistance
- **Uses:** fibreglass and carbon fibre composites for aerospace, automotive, and marine industries.

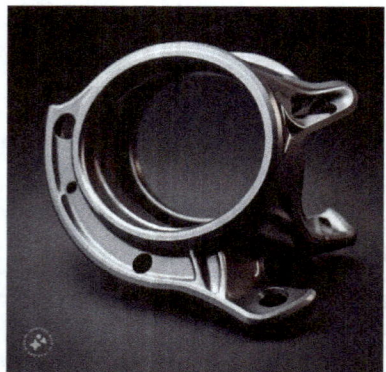
Automotive parts

marine parts

Thermoplastic and thermoset polymers are used to manufacture a variety of items. Both types offer distinct advantages and disadvantages, making them suitable for specific applications.

The choice of polymer type depends on several factors, including processing methods, mechanical properties, temperature resistance, chemical resistance, cost and the nature of the application

Chapter 3
Microplastics – Formation, Detection and Impacts

Microplastic

The term "**microplastics**" was first introduced in 2004 by Professor Richard Thompson from the United Kingdom. Microplastics are tiny plastic fragments that are less than 5 millimeters in size. They are found in various forms and are generally classified into two categories: **primary microplastics** and **secondary microplastics**.

Primary microplastics include items that are intentionally manufactured to be small, such as microbeads used in cosmetics, microfibers released from synthetic clothing during washing, and plastic pellets used in industrial production. On the other hand, **secondary microplastics** are formed when larger plastic products—like bags, bottles, and containers—break down over time due to exposure to sunlight, wind, water, and other environmental factors.

Many of us are familiar with the harmful effects of large plastic items through discussions in newspapers, on television, in YouTube videos, and across social media platforms. However, what often escapes our attention is the serious problem caused by tiny plastic particles that cannot even be seen with the naked eye.

To understand this better, let us consider a simple example from nature. When a living organism dies, it is gradually broken down by fungi and other decomposers until it eventually disappears from sight. Similarly, non-living materials, including plastic, also break down over time when they are left unused in the environment. But unlike organic matter, plastic does not fully disappear. Instead, it breaks into increasingly smaller pieces. These particles, though no longer visible, continue to exist in the environment. These are what we call **microplastics**.

While the invention of plastic once brought great convenience to human life, its harmful side has become increasingly visible in recent times. Among the most pressing concerns today is the widespread presence of microplastics, which pose serious risks to ecosystems, animals, and even human health.

Detection of plastic debris

In 1972, marine biologists Edward J. Carpenter and Kenneth L. Smith Jr. collected water samples from the Sargasso Sea in the North Atlantic Ocean to study surface water conditions.

They discovered large amounts of plastic pellets (pre-production plastic used to manufacture products) floating in the sea. In their published paper, they documented the presence of plastic pellets, fibres, and fragments on the ocean's surface and highlighted the potential impact of this debris on marine organisms, which were ingesting the particles. This was the first scientific documentation of plastic's persistence in marine environments.

Recently, Biochemist Janice Brahney, after analysing dust collected from the western United States, found that it contained a significant amount of fine plastic particles. These tiny particles are smaller than 5 millimetres in size.

Various reasons have been listed for the widespread presence of these fine plastic particles. They can originate from the breakdown of larger plastic items through various processes. Here's how this disintegration occurs:

Breakdown Processes Leading to Microplastics:

1. UV Radiation

Photo degradation: Exposure to sunlight causes plastics to break down chemically. Ultraviolet (UV) light weakens the chemical bonds in the plastic, leading to fragmentation into smaller pieces.

2. Mechanical Wear

Physical Abrasion: When plastics are subjected to friction, such as in washing machines or on roads, they can wear down and produce smaller particles. For example, synthetic fibres from clothing can shed during washing, and when vehicles travel on roads, their tires wear down, releasing tiny, often invisible plastic particles. These particles can mix with rainwater

3. Chemical Degradation

Chemical Reactions: Environmental factors like heat, humidity, and the presence of certain chemicals can break down plastics. This degradation can lead to the formation of microplastics.

4. Weathering

Environmental Factors: Exposure to wind, rain, and temperature changes can physically break down larger plastic items into smaller fragments over time.

Since plastics are made up of hydrocarbons, their breakdown can also release methane and ethane. Five major commodity thermoplastics commonly encountered in microplastics are Polyethylene, Polypropylene, Polyvinyl Chloride, Polystyrene and Polyethylene Terephthalate..

1. Low-Density Polyethylene (LDPE)

- **Recyclability**: Less commonly recycled, but can be recycled into products like garbage bags or plastic furniture.
- **Environmental Impact**: Contributes significantly to plastic pollution, especially in single-use plastic bags.

2. Polypropylene (PP)

- **Recyclability**: Can be recycled, but recycling rates are low compared to other plastics.
- **Environmental Impact**: Common in consumer products, contributing to landfill waste, though more resistant to breaking down into microplastics.

3. Polyvinyl Chloride (PVC)

- **Recyclability**: Difficult to recycle due to the release of harmful chemicals during recycling.
- **Environmental Impact**: Toxic if burned, and the production of PVC releases harmful chemicals.

4. Polystyrene (PS) or Expanded Polystyrene (EPS)

- **Recyclability**: Difficult to recycle due to its lightweight nature and contamination.
- **Environmental Impact**: Major contributor to pollution, particularly in oceans and landfills, where it can break into microplastics.

5. Polyethylene Terephthalate (PET or PETE)

- **Recyclability**: Highly recyclable, often recycled into fabric or new containers.
- **Environmental Impact**: Widely recycled, but PET in the environment can break down into microplastics

Rate of Disintegration of Plastic:

The rate of disintegration of plastic refers to the speed at which plastic materials break down into smaller particles or components over time due to environmental factors like sunlight, heat, moisture, and microbial activity. The most important factor influencing this rate is the chemical composition of the plastic, particularly the type of polymer used to create it. Plastics are generally known for their slow degradation rates. Below are some common estimates for how long different types of plastics take to break down.

Type of Plastic	Estimated Time to Disintegrate
Polylactic Acid (PLA)	3 to 6 months
Polypropylene (PP)	20 -30 years
Polyethylene Terephthalate (PET or PETE)	20-30 years
Nylon	30 to 40 years
Polystyrene (PS) or Expanded Polystyrene	50 -100 years

(EPS)	
Low-Density Polyethylene (LDPE)	50 -100 years
High-Density Polyethylene (HDPE)	100 to 500 years
Acrylonitrile Butadiene Styrene (ABS)	100 to 300
Polyvinyl Chloride (PVC)	100 -500 years
Polycarbonate (PC)	500 to 1,000 years

The Factors that Determine the Disintegration of Plastic:

1. Nature of Polymer

Plastics are made from long chains of molecules known as polymers, and the strength of the chemical bonds that hold these polymers together determines how durable the plastic will be. For instance, **polyethylene**, commonly used in plastic bags, consists of long, durable polymer chains that resist breaking down.

In contrast, **polystyrene**, often used in foam products like disposable cups, has a different molecular structure, making it more susceptible to degradation. As a result, polyethylene can take hundreds of years to break down, while polystyrene, under the right conditions, may disintegrate in a few decades.

2. Role of Additives

Plastics are rarely made from pure polymers. They often contain additives that enhance their properties, such as increasing flexibility, making them heat-resistant or providing colour. These additives can significantly alter the disintegration rate of plastic.

For example, stabilizers are often added to protect plastic from breaking down due to sunlight or heat. While these additives extend the useful life of plastic products, they also slow the degradation process once discarded. On the other hand, biodegradable plastics may contain additives that encourage microbial activity, speeding up disintegration under appropriate conditions.

3. Thickness and Density of the Plastic

The time it takes for plastic to break down is also affected by its physical structure, particularly thickness and density. Thicker and denser plastics contain more material to degrade, making the process slower. For example, a thick plastic bottle may take centuries to break down, while a thin plastic wrapper might disintegrate in just a few years. This difference occurs because thinner plastics have a larger surface area relative to their volume, allowing environmental factors like light and microbial activity to work more effectively on breaking down the material.

4. Environmental Conditions

The environment where plastic is disposed of plays a critical role in how long it takes to disintegrate. Plastics degrade at different rates depending on exposure to sunlight, oxygen, and microorganisms.

Plastics exposed to direct sunlight break down more quickly through photodegradation, where ultraviolet (UV) light from the sun breaks the chemical bonds in the plastic, making it brittle and causing it to fragment. In contrast, plastics buried in landfills, deprived of light and oxygen, may take hundreds of years to degrade.

In marine environments, plastic degradation can be even slower, as limited sunlight and cold temperatures hinder chemical reactions. Plastic waste in the ocean often breaks into smaller pieces, known as microplastics, which persist in the environment for decades and pose a threat to marine life and ecosystems.

5. Biodegradable and Compostable Plastics

In response to the growing plastic waste problem, manufacturers have developed **biodegradable** and **compostable plastics**. These materials are designed to break down much faster than traditional plastics, typically within months or a few years.

Biodegradable plastics contain additives that promote microbial activity, breaking the material into natural substances like water, carbon dioxide, and biomass. Compostable plastics go further by completely disintegrating in industrial composting facilities, leaving no toxic residue.

However, these plastics still require specific conditions, such as high temperatures and moisture, to degrade effectively. If disposed of improperly, they may persist in the environment for longer periods.

3.1 Recent Scientific Findings on Microplastic Pollution and Its Impact

1. Microplastics in Our Rain: An Unseen Storm

Imagine stepping outside after a gentle rain, feeling the cool droplets on your skin. But what if I told you that with every rainfall, something else-something invisible to the naked eye-is falling from the sky along with the water? Recent research has revealed a surprising and unsettling truth: our rain is carrying microplastics.

The Surprising Discovery

A team of scientists from the US Geological Survey (USGS) set out to study nitrogen pollution in rainwater. What they found was far more unexpected. Out of samples collected from eight different sites between Denver and Boulder, Colorado, over 90 percent contained plastic particles. Even more striking, urban areas showed higher concentrations than remote, mountainous regions.

But these weren't just any plastics. The samples sparkled with flecks of blue, red, silver, purple, and green-tiny fragments likely shed from synthetic clothing, industrial products, and everyday litter. The most common color? Blue, hinting at the widespread use of blue-dyed textiles and plastics in our daily lives.

Where Do These Microplastics Come From?

You might be wondering: how do these tiny plastic bits end up in the sky? The answer lies in our homes and industries. Every time we wash our clothes, especially those made from synthetic fabrics like polyester or blends, millions of microscopic fibers are released into the water. One study from the University of Plymouth found that a single load of laundry could send up to 728,789 fibers down the drain! This water eventually makes its way to rivers and oceans, carrying microplastics with it.

Factories also play a role, releasing plastic dust and fragments into the air. Over time, these particles can be swept up by the wind, traveling vast distances before eventually joining the clouds above us.

Plastic Rain: A New Environmental Challenge

These airborne microplastics don't just float around harmlessly. They mix with rainwater and fall back to earth as what scientists are now calling "plastic rain." This phenomenon doesn't just affect our water-it seeps into the soil, potentially impacting crops, wildlife, and even the food we eat.

How Microplastics Might Change the Weather

As if plastic rain wasn't concerning enough, scientists are now uncovering another twist in the tale: microplastics may actually be changing our weather.

Tiny Particles, Big Impact

Researchers at Pennsylvania State University have discovered that microplastics in the air can act as ice-nucleating particles. In simple terms, these tiny plastic flakes help form ice crystals in clouds-a role traditionally played by natural particles like dust and pollen.

Think of clouds as giant floating factories, where water droplets gather around particles, freeze, and eventually fall as rain or snow. When events like dust storms or volcanic eruptions fill the air with more particles, they can change how and when it rains. Now, microplastics are joining this atmospheric mix, present everywhere, all the time.

Delayed Rain, Sudden Downpours

So, what does this mean for the weather you experience? With more aerosols-including microplastics-in the sky, clouds can hold more water before releasing it. This can delay rainfall, but when it does come, it may be much heavier than usual. As Professor Miriam Freedman from Penn State explains:

"With more aerosols in the air, you collect more total water in the cloud before the droplets are large enough to fall, and so you get heavier rainfall when it comes."

Imagine longer dry spells, followed by sudden, intense downpours-an unsettling prospect for farmers, city planners, and anyone living in flood-prone areas.

A New Puzzle for Climate Scientists

The presence of microplastics in the atmosphere is a new and complex challenge for climate science. While these particles are currently less abundant than other man-made aerosols, they are lighter and can travel farther, lingering in the air for longer periods. Scientists are only beginning to understand just how much plastic is swirling above our heads-and what it means for our planet's future.

The Unfolding Story

The story of microplastics in our rain and clouds is just beginning. What started as a curiosity has become a critical environmental issue, with implications that reach from our laundry rooms to the very clouds above us. As researchers continue to unravel this mystery, one thing is clear: these tiny particles are having a much bigger impact than anyone imagined, adding yet another layer to the complex puzzle of environmental change.

So, the next time you watch the rain fall, remember-you're witnessing not just a weather event, but a chapter in the ongoing story of our changing planet.

2. Oceanic and Marine Ecosystems

Microplastics in the Gulf of İzmir: A Threat to Marine Life and Humans

A groundbreaking study has detected microplastics in varying amounts in the digestive systems of 11 commercial fish species in the Gulf of İzmir, Türkiye. The research, conducted by scientists from Recep Tayyip Erdoğan University, reveals that microplastics are pervasive in the marine ecosystem, posing a threat to both marine life and human health.

The Study's Findings:

The researchers examined the digestive systems of fish such as mackerel, red mullet, sea bream, and sardines, and found microplastics in amounts ranging from 101 to 4,901 micrometers.

The results showed that:

- 55% of mullet fish contained microplastics
- 50% of mackerel, 42% of horse mackerel, and 41% of bream also tested positive for microplastics
- The most common colors of microplastics were black (64.9%), followed by red, blue, green, white, and transparent

The Source of Microplastics

According to Kenan Gedik, the director of Recep Tayyip Erdoğan University's Vocational School of Technical Sciences, the source of microplastics is vast and varied. "All plastics, such as food packaging, bags, disposable materials, and synthetic textiles, contribute to microplastic pollution." Gedik noted that approximately 20 million tons of plastic waste are released into the seas every year, with most being single-use plastics.

The Impact on Marine Life

The study's findings suggest that microplastics can accumulate in pelagic fish species, which live near the water surface, more than in demersal fish, which live on the seabed.

Gedik explained that microplastics can block the digestive tracts of fish, causing problems with nutrient absorption, and even act as carriers for pollutants. This can lead to imbalances in the food chain and affect the entire marine ecosystem.

The Human Health Implications

Gedik warned that humans are also exposed to microplastics through the food chain, particularly when consuming fish contaminated with microplastics.

The chemicals added to plastics during production, such as phthalates and bisphenol A (BPA), can cause changes in liver function, insulin resistance, reproductive problems, and developmental disorders.

3. The Threat of Microplastics to Coral

What are Corals?

Corals are tiny, soft-bodied marine animals that play a vital role in supporting underwater life. Despite their small size, they're crucial for maintaining a healthy ocean ecosystem.

The Magic of Coral Reefs

Coral reefs are like underwater cities, built from the hard skeletons of corals made from calcium carbonate. These vibrant ecosystems provide a home, food, and shelter for a diverse range of sea creatures, earning them the nickname "rainforests of the sea." Coral reefs also act as natural barriers, shielding coastlines from big waves, storms, and erosion.

The Trouble with Corals and Coral Reefs

Unfortunately, corals and coral reefs are facing a grave threat. Recent research by scientists at National Sun Yat-sen University reveals that microplastics are damaging coral skeletons, weakening their structure and making them prone to damage. Key findings from the study include:

- Exposure to microplastics for just seven days caused significant damage to coral skeletons.
- Microplastics weakened the stable materials calcite and aragonite, leading to the formation of unstable amorphous calcium carbonate.
- This degradation altered calcium ion levels in surrounding seawater, potentially harming other marine life.

Why Coral Reefs Matter

Coral reefs are breathtakingly beautiful and essential for marine life. They protect coastal areas, support biodiversity, and provide food for millions. However, microplastic pollution is destroying these vital ecosystems, threatening marine life and the livelihoods of coastal communities.

4. Soil and Agricultural Impacts

i) Plastic Mulches: A Source of Microplastics

Plastic mulches are widely used in agriculture to retain moisture and suppress weeds. However, over time, they break down and release microplastics directly into the soil. This can lead to long-term consequences for soil health and plant growth.

Compost Contamination

Compost is a vital component of healthy farming, but synthetic materials like plastic mulches and fertilizers can contaminate it with microplastics. Even natural processes like composting can be affected by plastic pollution.

The Impact on Roots

Microplastics can accumulate around root hairs, hindering their ability to absorb water and nutrients. This disrupts root development, causing roots to grow abnormally and preventing plants from accessing essential resources. As microplastics clog the soil, they reduce its porosity, making it harder for plants to take up nutrients.

The Consequences for Crop Yields

The accumulation of microplastics in soil can reduce crop yields, leading to food insecurity. Plants may also absorb harmful substances like pesticides, posing risks to human health.

ii) Earthworms: Nature's Little Superheroes

Earthworms are often overlooked, but these soil dwellers play a critical role in boosting global food production and protecting plants from modern threats like microplastics.

The Impact of Earthworms on Food Production

According to a study by Colorado State University, earthworms contribute to the production of up to 140 million tons of food globally every year. By improving soil health, they enhance the growth of key crops like rice, maize, wheat, barley, and legumes such as soybeans and peas. In areas where farmers have limited access to fertilizers, earthworms act as natural soil enhancers.

How Earthworms Help

So, how do earthworms work their magic? Their constant tunneling loosens and aerates the soil, creating channels that allow water to drain effectively and nutrients to circulate. They also mix organic materials into the soil, making it rich and fertile for crops to thrive.

Earthworms vs. Microplastics

But earthworms don't stop there! A study by Japanese scientists revealed that these incredible creatures can protect plants from microplastic pollution.

In experiments, tomato plants grown in soil contaminated with microplastics thrived when earthworms were present. Not only did the plants grow healthier and stronger, but the earthworms also helped fend off pests. Scientists believe that earthworms enhance the plants' immune systems, making them more resilient to stress and disease.

The Importance of Earthworms

Earthworms are essential for sustainable agriculture. By managing soil in a more sustainable way, we can harness the power of these soil engineers to create greener ecosystems and healthier crops. Earthworms may be small, but their impact on our planet is enormous.

iii) Microplastic in Chlorophyll

The Hidden Threat of Microplastics: How Tiny Plastics Impact Plant Growth

Microplastics, tiny plastic pieces less than five millimeters in size, have become a pressing global environmental concern. These miniature pollutants are omnipresent, floating in the air, flowing through rivers and streams, and accumulating in soils. Recent research reveals a disturbing truth: microplastics are harming plant growth worldwide by significantly reducing photosynthesis rates.

A groundbreaking study published in the Proceedings of the National Academy of Sciences estimates that microplastics decrease photosynthesis by 7 to 12 percent. This decline could have far-reaching consequences for staple crops like wheat, corn, and rice. By analyzing data from 157 studies on plant responses to microplastics, researchers found consistent decreases in chlorophyll content, a key indicator of photosynthesis.

The numbers are staggering: microplastics might cause annual losses of 60 million tons of rice, 76 million tons of wheat, and 109 million tons of corn, representing a 9.6% reduction in global crop yields. However, experts emphasize that these estimates are preliminary and based on lab experiments rather than real-world conditions.

Several factors could influence the actual impact of microplastics on plant growth. The study's assumptions about microplastic shape and size might not reflect environmental realities. Moreover, these plastic particles can carry toxic substances like cadmium, further threatening plant health.

Despite these uncertainties, the study underscores the urgent need for more research to fully comprehend the effects of microplastics on ecosystems and agricultural productivity. As the world grapples with the challenges of microplastics, it's essential to explore the intricacies of this issue and its potential consequences for our planet's future.

v) Microplastics and Honey Bess: Threats to Pollinators

Imagine a honeybee hovering near a blooming lavender or rose. It stretches out its long, straw-like tongue to sip the sweet nectar, all while helping pollinate the flower. This is nature at work – beautifully efficient and vital for our crops and ecosystems.

But things start to fall apart when bees are exposed to microplastic pollution. New research shows that these tiny plastic particles can interfere with bees' memory. After exposure, bees may forget the scents linked with sugary rewards. If a bee forgets which flower smells like nectar, it may stop visiting it. As a result, pollination suffers – and that's bad news not just for flowers, but for the food we rely on.

Why Pollination Matters

Honeybees are the world's most important crop pollinators, but they're not the only ones. Bumblebees and other pollinators also play a role. As they move from flower to flower collecting nectar and pollen to feed themselves and their young, they also help plants reproduce by transferring pollen. This simple action is essential for the growth of fruits, vegetables, and seeds.

How Microplastics Enter the Picture

Microplastics are tiny plastic particles, often less than 0.2 inches in size. They come from things like food packaging, synthetic clothing, disposable cutlery, and even glitter and body scrubs. Some are already small, while others break down from larger plastic waste through exposure to sunlight, wind, and water.

These microplastics can be found everywhere – in the air, in soil, on plants, and in water. When wastewater containing microplastics is used to irrigate crops, or when plastic waste breaks down in nature, bees can easily inhale or ingest these particles while flying or feeding.

Bees at Risk

Once inside a bee's body, microplastics can do serious damage. Studies have shown they can:

- Damage the bee's gut and even reach its brain.
- Affect memory and learning – bees exposed to microplastics forget trained scent associations within just 24 hours.
- Increase vulnerability to harmful bacteria and viruses.
- Lead to physical changes like weight loss, hair loss, or darker body color.
- Even cause death. One study found that microplastics increased bee mortality by up to 25%.

Some microplastics can also stick to a bee's body – their wings, head, and abdomen – or be transported to flowers, where they may block essential parts like the stigma. If pollen can't reach the stigma, fertilization fails, and seeds don't form.

Plastic in Honey, Flowers, and Soil

The presence of microplastics isn't limited to bees – it also ends up in honey. Studies in Turkey and Germany have found hundreds of microplastic particles in honey samples, especially in single-flower varieties.

Flowers and plants suffer too. In a 2024 study, scientists found that plastic particles jammed the stigma of Andean yellow monkeyflowers, preventing successful pollination. Tomato plants grown in plastic-polluted soil produced 28% fewer flowers, and similar effects were observed in mustard plants. In general, plants growing in such soil tend to be smaller and weaker.

What This Means for Us

Microplastics don't just pollute the oceans; they're now deeply embedded in our land and agriculture. And the damage they do – from harming pollinators to reducing crop growth – could directly impact our food supply.

As David Baracchi, a behavioral ecologist from the University of Florence, says, "We are only beginning to scratch the surface" of understanding how microplastics affect pollinators.

And as Thomas Cherico Wanger-Guerrero, an agroecologist from Switzerland's federal research center, warns: "If plastic is adding to all the stressors that pollinators are facing already, I think we really might be in a tricky position."

The urgency is clear: we need to reduce plastic pollution not just in our oceans but also on farms, in our air, and across our ecosystems. Protecting pollinators means protecting the future of our food.

5. Microplastics and Microbes: A Dangerous Alliance

The Hidden Dangers of Microplastics: A Growing Threat to Human Health

You might know that germs like viruses and bacteria cause many diseases. But have you considered plastic's role in strengthening these germs? Research reveals a disturbing link between microplastics and severe health issues like heart disease, dementia, and cancer.

Microplastics, which break off from everyday items like shopping bags and food packaging, persist in the environment for up to 500 years. They contaminate our food chain, drinking water, and even our bodies.

A University of Oxford study found that microplastics directly contribute to the rise and spread of drug-resistant superbugs. These superbugs can fight off life-saving antibiotics, making them increasingly deadly.

The research suggests microplastics increase the spread of superbugs by as much as 200 times. If left unchecked, this could lead to millions of deaths. The World Health Organization warns that by 2050, ten million people may die annually due to superbugs.

Professor Timothy Walsh, a renowned microbiologist with over 25 years of experience, expresses concern: "Given the lack of global plastic waste governance and the increasing amount of microplastics infiltrating all aspects of human activity, these findings are very concerning."

6. The Silent Invasion: Microplastics in Human Organs

1. Microplastics and Heart Health: A Hidden Risk

A groundbreaking 2024 study published in the *New England Journal of Medicine* has sent shockwaves through the medical community. Researchers analyzed artery plaque from 257 patients undergoing procedures to reduce blockages and discovered microplastics in over half the cases. Even more alarming, these patients were **four times more likely** to experience a heart attack, stroke, or death within three years compared to those without detectable microplastics.

Why Are Microplastics So Dangerous?

The greatest threat comes from particles smaller than 1 micrometre, which can slip through the gut or lungs and enter the bloodstream. Larger particles, over 10 micrometres, are typically filtered out by the body's natural defenses. Laboratory studies have shown that microplastics can damage cells, tissues, and DNA, and may even promote cancer growth.

A Complex Mix of Plastics

More than a dozen types of plastics-ranging from jagged-edged specks to microscopic fiber strings-have been detected in the human body. Research from King's College London found that daily microplastic deposits in central London reached 1,000 particles per square meter, even when only particles larger than 20 micrometres were counted.

What Does This Mean for Our Health?

Although the full impact of microplastics on human health is still being uncovered, the evidence points to significant risks. Microplastics can infiltrate tissues, disrupt biological processes, and act as carriers for harmful chemicals. Ongoing research is crucial to fully understand their long-term effects.

ii) Microplastics in the Bloodstream: Plastic on the Move

A recent study highlighted in *The Weekly Voice* (December 2024) has uncovered a startling connection between microplastics in human blood and potential health risks. Researchers from South Korea examined blood samples from 36 healthy adults and found microplastics in nearly **89% of participants**, with an average of 4.2 particles per milliliter of blood.

Where Are These Plastics Coming From?

The most common types detected were polystyrene (used in disposable cups and containers) and polypropylene (found in food packaging). The study found that people who frequently use plastic food containers had significantly higher levels of microplastics in their blood.

How Do Microplastics Affect Blood Health?

Higher levels of microplastics were linked to noticeable changes in blood clotting markers, such as fibrinogen-a protein crucial for forming clots. While scientists are still investigating the exact health implications, these findings raise serious concerns about the potential for increased risks of blood clots and heart disease.

iii) Microplastics Invading Vital Organs: Brain, Liver, and Kidney

The Silent Threat in Our Brains

A pioneering study by the University of New Mexico analyzed brain, liver, and kidney samples from individuals who died between 1997 and 2024. The results were eye-opening: there has been a significant rise in micro- and nanoplastics in brain tissue over the years, with polyethylene-the plastic used in bags and food packaging-being the most common.

Microplastics and Neurological Health

Perhaps most concerning, microplastic concentrations were found to be **six times higher** in brain samples from individuals with dementia. While the study does not claim that microplastics cause dementia, their presence in the brain raises troubling questions about potential impacts on neurological health.

A Growing Trend

Further analysis of tissue samples from the U.S. East Coast between 1997 and 2013 showed a clear, increasing trend in microplastic contamination. The message is clear: our growing reliance on plastics is not just an environmental issue-it's a direct threat to our health, affecting our most vital organs.

vi) Microplastics in Human Placentas

Measuring Microplastics in Placentas

Researchers from the University of New Mexico Health Sciences have pioneered a new analytical tool to measure microplastics in human placentas-the organ that nourishes and protects a developing baby during pregnancy. Analyzing 62 placenta samples, they found microplastic concentrations ranging from 6.5 to 790 micrograms per gram of tissue. While these amounts may seem tiny, the implications for health are significant.

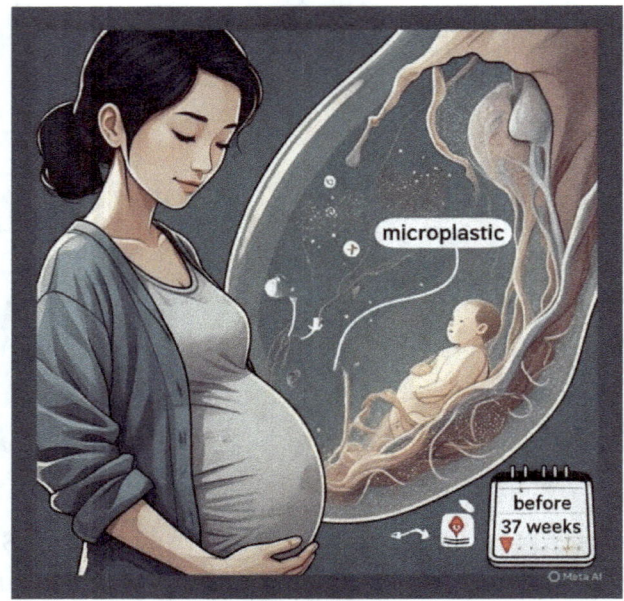

Types and Sources of Microplastics Detected

The study identified polyethylene-the plastic commonly used in bags and food packaging-as the most prevalent type (54%). Polyvinyl chloride (PVC) and nylon each made up around 10%, with other plastics comprising the remaining 26%. These findings highlight how everyday plastic products contribute to microplastic accumulation in the body, and demonstrate the challenges scientists have faced in quantifying these particles until now.

Microplastics and Preterm Birth

Of particular concern is the discovery that preterm babies-those born before 37 weeks-have higher levels of microplastics in their placentas compared to full-term babies. Preterm birth is a leading cause of death in children under five and is associated with serious complications such as breathing and feeding difficulties. The study suggests that higher microplastic concentrations in preterm placentas may be linked to an increased risk of preterm labor, raising important questions about the role of environmental exposure in pregnancy outcomes.

Health Risks of Microplastics in Placental Tissue

Microplastics can enter the human body through food, water, and air. The smallest particles (less than 1 micrometre) can cross cell membranes and enter the bloodstream, while larger ones are usually expelled. Laboratory studies have shown that microplastics can damage cells, tissues, and DNA, and may even promote cancer growth. There are also concerns about links to inflammatory bowel disease and declining sperm counts.

The Growing Challenge of Plastic Exposure

Toxicologists warn that increasing exposure to microplastics is a growing concern for all mammalian life. The concentration of microplastics in placental tissue is especially troubling, as this organ develops rapidly during pregnancy and is crucial for fetal health. Despite these concerns, global plastic production is projected to double every 10 to 15 years, further compounding the problem.

Implications for Expectant Mothers

For expectant mothers, these findings underscore the importance of minimizing contact with microplastics whenever possible. While much remains to be learned about the full health impacts, this research paints a stark picture: our reliance on plastics is not only harming the planet but may also be affecting the very beginning of human life.

7. Microplastics in IV fluids

Recent research from Fudan University in Shanghai has raised alarms about the presence of microplastics in IV fluids used in hospitals. These tiny plastic fragments, resulting from the degradation of plastic materials, have been found in various parts of the human body, from the brain to the liver, heart, and even breast milk. Microplastics have been linked to serious health conditions, including cancer, heart disease, and inflammatory bowel diseases.

Microplastics in IV Bags: A Concerning Discovery

Researchers tested common IV saline solution bags and found that they release thousands of microplastic particles into the solutions. The numbers are staggering: approximately 7,500 microplastic particles in an 8.4 oz IV bag, 25,000 particles in a standard IV drip for dehydration, and over 52,000 particles for complex surgeries like abdominal procedures.

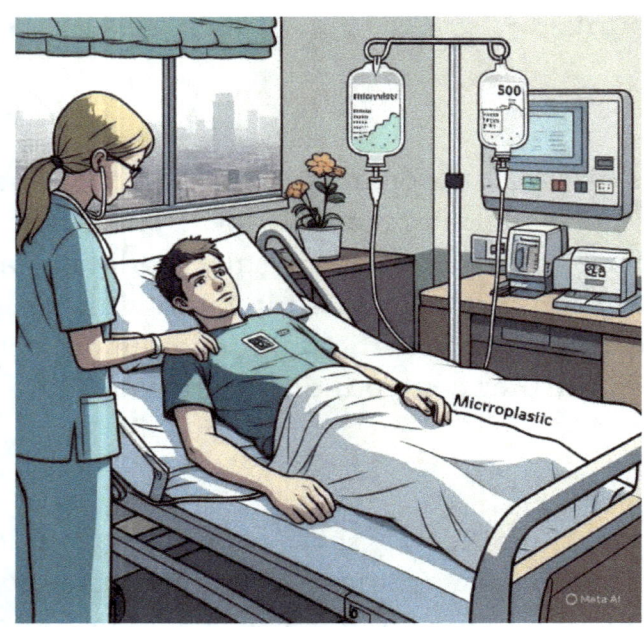

Perspective on the Risk

While these findings might seem alarming, experts emphasize that patients should not avoid IV treatments due to the potential risks posed by microplastics. The conditions treated with IVs, such as dehydration or surgeries, are far more dangerous than the potential risks associated with microplastics.

Reducing Exposure

Researchers do offer some solutions to minimize exposure. Keeping IV bags away from direct light or heat can reduce microplastic shedding. In the future, IV drips could include filters to remove most of the harmful particles.

8. Microplastics in tea bag

If you're a fan of loose-leaf tea, you might be onto something. Recent research from scientists in Spain reveals that commercial tea bags can leach millions of nanoplastics and microplastics into your cup. This study, published in the journal Chemosphere, highlights the potential risks of plastic pollution in our daily lives.

The Study's Findings

A team from the Autonomous University of Barcelona, Sohag University, and the Helmholtz Centre for Environmental Research discovered that polymer-based tea bags release enormous amounts of microplastics and nanoplastics when infused. The researchers found that these particles can be absorbed by intestinal cells, entering the bloodstream and potentially affecting the entire human body.

The Biological Impact

The study showed that mucus-producing intestinal cells are particularly susceptible to microplastic uptake, with particles even entering the cell nucleus. This raises concerns about the potential health implications of chronic exposure to environmental microplastics. The researchers emphasize the need for further investigation into the effects of microplastics on human health.

A Growing Concern

This study builds upon previous research that demonstrated the ability of plastic particles to enter the brain, testicles, and even pass from mother to unborn child. The findings also suggest that microplastics can interfere with antibiotic effectiveness. As we continue to uncover the extent of microplastic pollution, it's essential to consider the potential consequences for our health and well-being.

9. Invisible Risks: Microplastic Release from Non-Stick Cookware

Ditch the Non-Stick: A Healthier Kitchen Awaits

You can start reducing plastic consumption right from your home, and it begins with a common kitchen item: non-stick cookware. But have you ever wondered what's behind its convenience?

The Dark Side of Non-Stick

Non-stick pans are coated with polytetrafluoroethylene (Teflon). When heated, it releases harmful chemicals, and scratched coatings can leach microplastics into your food. Do you want to serve your loved ones, especially kids, food contaminated with toxic substances?

The Risks of Scratched Non-Stick Coatings

However, there's a catch. Non-stick cookware is vulnerable to scratches, and when it gets scratched, Teflon (PTFE) can start to break down. These chemicals are linked to health risks such as hormone disruption and, in extreme cases, can even release toxic fumes when heated to high temperatures.

What Happens When Teflon Coating is Damaged?

When the Teflon coating on your pan is scratched, microplastics and chemicals can seep into your food. These particles, when ingested, may affect your health over time. And if you're cooking at high heat, these coatings can release dangerous fumes that may be harmful to both humans and pets.

"*Microplastics are not just a marine issue; they can affect terrestrial ecosystems and human health, revealing the interconnectedness of our environment.*" - Dr. Richard Thompson, Professor of Marine Biology, Director of the Marine Institute, University of Plymouth

3.2 MICROBEADS

Microplastics are typically the result of the breakdown of larger plastic items due to environmental factors like sunlight, wind, and wave action. In contrast, microbeads are intentionally manufactured small plastic particles, often added to cosmetics and personal care products for their exfoliating properties. Microbeads are a subset of microplastics, characterized by their spherical shape and smaller size, typically ranging from 0.1 to 1 millimetre in diameter.

This distinction highlights two different sources of plastic pollution:

1. **Microplastics**: Accidental byproducts from the degradation of larger plastic debris.
2. **Microbeads**: Deliberately produced particles, raising concerns about their environmental impact despite their intended use.

Both contribute to the overall issue of plastic pollution, but their origins and the implications for regulation and public awareness differ.

Microbeads are manufactured to be precise and uniform in size (typically less than 1mm) and spherical in shape. This uniformity makes them effective in applications where consistency is essential, such as in exfoliants or polishing agents.

Microbeads are primarily used in the following industries:

1. Personal Care and Cosmetics Industry

- **Application:** Exfoliation and Skin Care
- **Products:** Face scrubs, body washes, toothpaste, shampoos, and cleansers
- **Reason for Use:**
  - Microbeads are used for exfoliation to gently remove dead skin cells.
  - Their smooth, round shape ensures a uniform texture and less irritation compared to harsher natural abrasives like walnut shells or pumice.
  - Microbeads also help in giving products a luxurious feel and uniform consistency.

2. Industrial Cleaning Products

- **Application:** Polishing and Abrasives
- **Products:** Automotive polishes, industrial surface cleaners, and specialized cleaning agents
- **Reason for Use:**
o Microbeads act as fine polishing agents, smoothing and shining surfaces like car exteriors or industrial materials.

3. Water and Air Filtration Systems

- **Application:** Filtration Media
- **Products:** Water filters, air filtration systems, and purification technologies
- **Reason for Use:**

o Microbeads are used as filtration media because their uniform size and shape create porous layers that trap particles and contaminants efficiently.
o Their chemically inert nature ensures that they do not react with the substances being filtered.

4. Paints, Coatings and Plastics Industry

- **Application:** Lightweight Filler Materials
- **Products:** Paints, coatings, plastics and sealants
- **Reason for Use:**
  - They provide a smooth, even texture in coatings and help reduce production costs.

5. Textile and Fabric Care Industry

- **Application:** Detergents and Fabric Conditioners
- **Products:** Laundry detergents, fabric softeners, and stain removers
- **Reason for Use:**
  - Microbeads are included in detergents for their ability to enhance cleaning power, acting as small abrasives that help lift dirt and stains from fabrics.
  - They also improve the flowability of powdered detergents, preventing clumping and ensuring uniform distribution during the wash.

6. Automotive and Aerospace Industry

- **Application:** Surface Polishing and Smoothing
- **Products:** Car waxes, aerospace coatings, and surface protectants
- **Reason for Use:**
  - Microbeads are used for surface polishing in automotive and aerospace industries to create smooth, shiny surfaces..

7. Household Cleaning Products

- **Application:** Scrubbing Agents
- **Products:** Kitchen cleaners, bathroom scrubs, and multi-surface cleaners
- **Reason for Use:**
  - Microbeads serve as scrubbing agents that help remove tough dirt, grime, and stains from surfaces like countertops, tiles, and sinks.
  - Their abrasive yet gentle properties make them effective in cleaning without scratching surfaces.

Microbeads, commonly found in personal care products, household cleaners, and paints, present significant environmental concerns primarily due to their disposal and degradation. When these products are rinsed off, microbeads are sent directly into wastewater systems, often bypassing treatment processes and entering waterways, contributing to pollution. Their small size allows them to escape filtration systems, leading to contamination of natural water sources.

Over time, microbeads in paints and coatings can degrade, releasing microplastics into the environment, further exacerbating the problem. Additionally, microbeads from detergents and polishing agents can escape wastewater treatment systems, resulting in plastic pollution in rivers and oceans.

Even though microbeads have been banned in many countries, we still need to be more conscious when buying these products in the market. The most common types of plastics used for manufacturing microbeads include:

- *Polyethylene (PE)*
- *Polypropylene (PP)*
- *Poly methyl methacrylate (PMMA)*
- *Nylon*
- *Polyethylene terephthalate (PET)*
- *Acrylates copolymer*
- *Polystyrene (PS)*

Hence, while purchasing products, check the ingredient list on the packaging for the above plastic-based ingredients.

Chapter 4
Global Perspectives and Legislative Efforts

1. World Health Organisation Concerns Over Drinking Water

The World Health Organization (WHO) has expressed concern over microplastics, calling for further assessment of their presence in the environment and their potential impacts on human health. This follows the release of an analysis of current research related to microplastics in drinking water. The organization also advocates for a reduction in plastic pollution to benefit the environment and reduce human exposure.

"We urgently need to know more about the health impact of microplastics because they are everywhere - including in our drinking-water," says Dr Maria Neira, Director, Department of Public Health, Environment and Social Determinants of Health, at WHO. "Based on the limited information we have, microplastics in drinking water don't appear to pose a health risk at current levels. But we need to find out more. We also need to stop the rise in plastic pollution worldwide."

According to the analysis, which summarizes the latest knowledge on microplastics in drinking-water, microplastics larger than 150 micrometres are not likely to be absorbed in the human body and uptake of smaller particles is expected to be limited. Absorption and distribution of very small microplastic particles including in the nano size range may, however, be higher, although the data is extremely limited.

Further research is needed to obtain a more accurate assessment of exposure to microplastics and their potential impacts on human health. These include developing standard methods for measuring microplastic particles in water; more studies on the sources and occurrence of microplastics in fresh water; and the efficacy of different treatment processes.

WHO recommends drinking-water suppliers and regulators prioritize removing microbial pathogens and chemicals that are known risks to human health, such as those causing deadly diarrhoeal diseases. This has a double advantage: wastewater and drinking-water treatment systems that treat faecal content and chemicals are also effective in removing microplastics.

Wastewater treatment can remove more than 90% of microplastics from wastewater, with the highest removal coming from tertiary treatment such as filtration. Conventional drinking-water treatment can remove particles smaller than a micrometre.

A significant proportion of the global population currently does not benefit from adequate water and sewage treatment. By addressing the problem of human exposure to faecally contaminated water, communities can simultaneously address the concern related to microplastics.

2. European Commission's Press Release on Restricting Microplastics:

In a press release on 25.09.2023, the European Commission stated that it is taking another major step to protect the environment by adopting measures to restrict intentionally added microplastics in products under the EU chemical legislation, REACH.

Further, it states it will prevent the release to the environment of about half a million tonnes of microplastics. They will prohibit the sale of microplastics as such, and of products to which microplastics have been added on purpose and that release those microplastics when used. When duly justified, derogations and transition periods for the affected parties to adjust to the new rules apply.

The adopted restriction uses a broad definition of microplastics – it covers all synthetic polymer particles below five millimetres that are organic, insoluble and resist degradation. The purpose is to reduce emissions of intentional microplastics from as many products as possible. Some examples of common products in the scope of the restriction are:

- The granular infill material used on artificial sport surfaces – the largest source of intentional microplastics in the environment;
- Cosmetics, where microplastics is used for multiple purposes, such as exfoliation (microbeads) or obtaining a specific texture, fragrance or colour;
- Detergents, fabric softeners, glitter, fertilisers, plant protection products, toys, medicines and medical devices, just to name a few.

Products used at industrial sites or not releasing microplastics during use are derogated from the sale ban, but their manufacturers will have to provide instructions on how to use and dispose of the product to prevent microplastics emissions.

Next Steps

The first measures, for example the ban on loose glitter and micro*beads*, will start applying when the restriction enters into force in 20 days. In other cases, the sales ban will apply after a longer period to give affected stakeholders the time to develop and switch to alternatives.

Background

The Commission is committed to fighting microplastics pollution, as stated in the European Green Deal and the new Circular Economy Action Plan. In the Zero Pollution Action Plan, the Commission set the target to reduce microplastics pollution by 30% by 2030.

As part of these efforts, the Commission is working to reduce microplastics pollution from different sources: plastic waste and litter, accidental and unintentional releases (e.g. plastic pellet loss, tyres degradation or release from clothing), as well as intentional uses in products.

To tackle microplastics pollution while preventing the risk of fragmentation in the single market, the Commission requested the European Chemicals Agency (ECHA) to assess the risk posed by microplastics intentionally added to products and whether further regulatory action at EU level was needed. ECHA concluded that microplastics intentionally added to certain products are released into the environment in an uncontrolled manner, and recommended to restrict them.

Based on the scientific evidence provided by ECHA, the Commission drafted a restriction proposal under REACH that was positively voted by the EU countries and successfully passed the scrutiny of the European Parliament and the Council before being adopted.

3. World Economic Forum's Warning on Microplastic

The World Economic Forum (WEF) is an international organization headquartered in Geneva, Switzerland. a non-governmental organization (NGO), in its latest update says Microplastics have been found in the land, sea and air across the food chain and throughout the human body.

It is still unclear what the environmental and health impacts of microplastics could be, but a recent study point to the possibility that they can increase the likelihood of heart attack, stroke or death. It is estimated that the average person can eat drink or breathe between 78,000 and 211000 microplastic particles every year.

In March 2022, at the UN Environment Assembly 175 nations agreed to end plastic pollution. A legally binding agreement – addressing the issues of single use plastic and recycling technology, among others – will be drawn up by the end of 2024.

4. United Nations Environment Programme (UNEP) Report on Microbeads:

Microbeads are a type of microplastic used in a variety of products. Their size can range from 1 micrometre to 1000 micrometres, and they are sometimes too small to be seen by the naked eye.

These microbeads are made from materials like polyethylene and they are used in many personal care and cosmetic products (PCCPs) like face washes, toothpaste, and makeup.

Microbeads are added to products like scrubs and toothpaste to help gently scrub away dead skin or plaque. They help make the skin or teeth feel smooth.

Plastics can help make products like shampoo and lotion feel smooth and creamy when applied to the skin or hair.

In some products like makeup or sunscreen, plastics help create a thin layer on the skin to lock in moisture and make the product last longer, especially in water-resistant formulas.

As a result, many countries are now banning or regulating the use of microplastics in PCCPs, encouraging companies to replace them with natural alternatives like ground nutshells or salts

Plastics can also be used to bulk up powders or makeup without changing how the product works.

5. Food and Agriculture Organisation of the United Nations's Report

A new report by the Food and Agriculture Organization of the United Nations (FAO), published on December 7, 2021, suggests that the land we use to grow our food is contaminated with far larger quantities of plastic pollution, posing an even greater threat to food security, people's health, and the environment.

The report - "Assessment of agricultural plastics and their sustainability: a call for action" - is the first global report of its kind by FAO and contains some startling number.

According to data collated by the agency's experts, agricultural value chains each year use 12.5 million tonnes of plastic products. A further 37.3 million tonnes are used in food packaging. The crop production and livestock sectors were found to be the largest users, accounting for 10.2 million tonnes per year collectively, followed by fisheries and aquaculture with 2.1 million tonnes, and forestry with 0.2 million tonnes.

Asia was estimated to be the largest user of plastics in agricultural production, accounting for almost half of global usage. In the absence of viable alternatives, demand for plastic in agriculture is only set to increase.

According to industry experts, for instance, global demand for greenhouse, mulching and silage films will increase by 50 percent, from 6.1 million tonnes in 2018 to 9.5 million tonnes in 2030.

Such trends make it essential to balance the costs and benefits of plastic. Of increasing concern are microplastics, which have the potential of adversely affecting human health. While there are gaps in the data, they shouldn't be used as an excuse not to act, FAO warned.

"This report serves as a loud call to coordinated and decisive action to facilitate good management practices and curb the disastrous use of plastics across the agricultural sectors," FAO Deputy Director-General Maria Helena Semedo said in the report's forward.

The Good

Plastics have become ubiquitous since their widespread introduction in the 1950s, and it is difficult today to envisage life without them.

In agriculture, plastic products greatly help productivity. Mulch films, for instance, are used to cover the soil to reduce weed growth, the need for pesticides, fertilizer and irrigation; tunnel and greenhouse films and nets protect and boost plant growth, extend cropping seasons and increase yields; coatings on fertilizers, pesticides and seeds control the rate of release of chemicals or improve germination; tree guards protect young seedlings and saplings against damage by animals and provide a microclimate that enhances growth.

Moreover, plastic products help reduce food losses and waste, and maintain its nutritional qualities throughout a myriad of value chains, thereby improving food security and reducing greenhouse gas (GHG) emissions.

The bad and the ugly

Unfortunately, the very properties that make plastics so useful create problems when they reach the end of their intended lives.

The diversity of polymers and additives blended into plastics make their sorting and recycling more difficult. Being man-made, there are few microorganisms capable of degrading polymers, meaning that once in the environment, they may fragment and remain there for decades. Of the estimated 6.3 billion tonnes of plastics produced up to 2015, almost 80 percent has not been disposed of properly.

Once in the natural environment, plastics can cause harm in several ways. The effects of large plastic items on marine fauna have been well documented. However, as these plastics begin to disintegrate and degrade, their impacts begin to be exerted at the cellular level, affecting not only individual organisms but also, potentially, entire ecosystems.

Microplastics (plastics less than 5 mm in size) are thought to present specific risks to animal health, but recent studies have detected traces of microplastic particles in human faeces and placentas. There is also evidence of mother-to-foetus transmission of much smaller nanoplastics in rats.

While most scientific research on plastics pollution has been directed at aquatic ecosystems, especially oceans, FAO experts found that agricultural soils are thought to receive far greater quantities of microplastics. Since 93 percent of global agricultural activities take place on land, there is an obvious need for further investigation in this area.

Key recommendations

The absence of viable alternatives makes it impossible for plastics to be banned. And there are no silver bullets for eliminating their drawbacks.

Instead, the report identifies several solutions based on the 6R model (Refuse, Redesign, Reduce, Reuse, Recycle, and Recover). Agricultural plastic products identified as having a high potential for environmental harm that should be targeted as a matter of priority include non-biodegradable polymer coated fertilizers and mulching films. The report also recommends developing a comprehensive voluntary code of conduct to cover all aspects of plastics throughout agrifood value chains and calls for more research, especially on the health impact of micro and nanoplastics.

"FAO will continue to play an important role in dealing with the issue of agricultural plastics holistically within the context of food security, nutrition, food safety, biodiversity and sustainable agriculture," Semedo said.

Pathways of Microplastics Contamination

Contaminated Seafood

One of the most well-known ways microplastics enter the human digestive system is through seafood. Microplastics are often released into oceans, rivers, and lakes from plastic waste that breaks down over time. Once in the water, they are mistaken for food by marine organisms such as fish, shellfish, and plankton. These marine animals ingest the particles, which then accumulate in their bodies.

When humans consume seafood, especially whole organisms like shellfish, the microplastics that have accumulated in their tissues are transferred to our bodies.

Studies have found microplastics in a wide variety of seafood products, including mussels, oysters, and fish commonly found in markets and restaurants. As seafood is a significant part of many diets globally, this pathway is a major source of microplastic ingestion.

Microplastics in Drinking Water

Another significant route through which microplastics indirectly enter our stomachs is via drinking water. Both bottled and tap water have been found to contain microplastic particles. These particles originate from the degradation of larger plastic items, which leach into water sources over time. In bottled water, the microplastics may come from the packaging itself, while in tap water, they can enter through pollution of natural water bodies.

In 2018, the non profit organization **Orb Media** conducted a survey of tap water from six regions on five continents and reported that, of the 159 samples analysed, 83% contained plastic particles.

Most of these particles were fibres (99.7%), measuring between 0.1mm and 5 mm in length. The number of particles ranged from 0 to 57 per litre, with an overall mean of 4.34 particles per litre.

The highest density of plastic per volume of tap water was found in North America, while the lowest densities were collectively found in seven European countries. The average person consuming water daily may ingest thousands of microplastic particles each year, contributing to their overall exposure

Food Packaging

Food packaging is another indirect source of microplastics. Many food items are wrapped or stored in plastic materials, which can degrade and release microplastics into the food they contain. This is especially true when plastic containers are heated, worn, or subjected to stress, as microplastic particles can leach out into the food.

Takeaway containers, plastic bottles, and cling film are all common sources of microplastic contamination. Heating food in plastic containers in the microwave, for instance, can increase the leaching of microplastics, leading to higher ingestion levels. Over time, small particles from these plastics may end up in our meals, even if they are not visible to the naked eye.

Agricultural Products

Microplastics can also make their way into agricultural products, further contributing to their indirect entry into the human body. Plastics used in agriculture—such as mulching films, irrigation pipes, and greenhouses—can break down into microplastic particles.

When fields are irrigated with microplastic-contaminated water or fertilized with sludge containing plastic waste, these particles can infiltrate the soil and be absorbed by crops. Microplastics that enter the body through ingestion and respiration can accumulate in various organs, including the lungs, intestines, kidneys, urinary tract, and liver.

According to studies, nearly 50,000 plastic particles enter our bodies each year through food and respiration. Larger plastic particles are expelled through feces, while smaller particles remain trapped inside the body.

The full extent of their impact is still not completely understood. However, plastic is a foreign substance that does not belong in the human body, and if it continues to accumulate, it can affect cells and impair the immune system.

Moreover, while the enzymes in our stomach help digest fats and proteins in the food we consume, there are no enzymes capable of digesting plastic particles.

Chapter 5
Plastics in Emerging Industries

Despite the pressing environmental concerns associated with plastic pollution, its significance in modern technologies remains undeniable. Over a century after its invention, plastic continues to play a pivotal role in various sectors, driving innovation and efficiency. Since its creation in the early 20th century, plastic has transformed countless industries, as we have already seen in the beginning of this book.

Here's an overview of how plastics support modern industries, such as artificial intelligence, electric vehicles, and telecommunications.

1. Artificial Intelligence (AI)

- Plastics are used in the manufacturing of lightweight and durable housing for AI devices, sensors, and robotics, allowing for innovative designs and portability. Components like circuit boards often incorporate plastic for insulation and protection.
- **Without Plastic:** The industry would rely more on metals and glass, increasing the weight and bulk of devices, complicating integration into everyday technology, and potentially hindering advancements in robotics due to design limitations.

2. Cyber security

- Many cyber security devices, like firewalls and routers, use plastic housings for durability and lightweight design, facilitating easy

installation in various environments. Plastic connectors and insulators help maintain device integrity.
- **Without Plastic**: Devices would be heavier and more fragile, leading to potential reliability issues. The design flexibility would be significantly reduced, making it harder to develop compact and efficient hardware solutions.

3. Electric Vehicles (EVs)

- Plastics contribute to weight reduction in EVs, improving efficiency and range. They are used in components like dashboards, battery casings, and electrical insulation, enhancing performance and safety.
- **Without Plastic**: EVs would be heavier and less efficient, increasing energy consumption. The design and production of lightweight components would be limited, leading to potential setbacks in performance and consumer adoption.

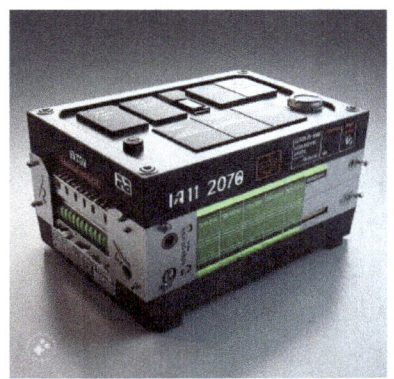 

4. Renewable Energy

- Plastics are used in solar panel manufacturing, wind turbine components, and energy storage systems, providing lightweight, durable materials that withstand environmental conditions.
- **Without Plastic**: Renewable energy technologies would be bulkier and more challenging to install and maintain. The absence of lightweight materials could slow the adoption of renewable energy solutions.

5. Block chain and Crypto currency

- Plastic is used in the production of hardware wallets, mining rigs, and data storage devices, providing lightweight and durable options that support secure transactions and data management.
- **Without Plastic**: The reliance on heavier materials could hinder the development of portable and secure devices. The overall efficiency of hardware used in crypto currency mining and storage would likely decrease.

6. Virtual and Augmented Reality (VR/AR)

- Virtual Reality and Augmented Reality headsets utilize plastics for lightweight designs, comfort, and durability, enhancing user experience. Plastic lenses and components are essential for functionality.
- **Without Plastic**: Virtual Reality and Augmented Reality devices would be heavier and less comfortable, potentially limiting user engagement and making technology less accessible. Design innovation would be stifled.

7. 3D Printing

- Plastics are a primary material in 3D printing, allowing for rapid prototyping and production of custom parts across various industries. The availability of different types of plastics enables a wide range of applications.
- **Without Plastic**: The 3D printing industry would be constrained to materials like metals and ceramics, which are more expensive and harder to work with, significantly reducing accessibility and innovation.

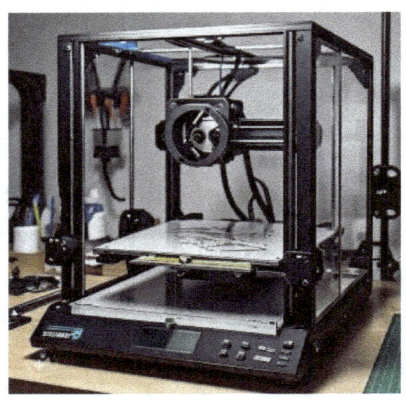

8, Biotechnology

- Plastics are used in lab equipment, disposable medical devices, and packaging for biopharmaceuticals, providing cost-effective and sterile solutions that support advancements in healthcare.
- **Without Plastic**: The industry would face challenges in maintaining sterility and safety in medical devices. The increased weight and cost of alternatives could hinder access to necessary tools and supplies.

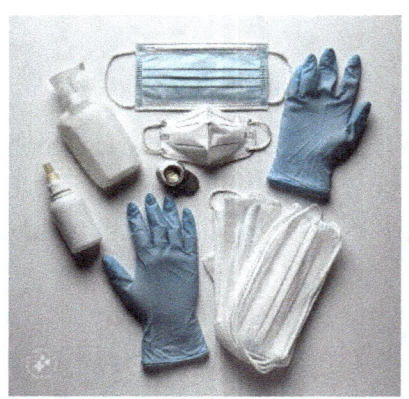

9. Nanotechnology

- Plastics can be engineered at the nanoscale for applications in drug delivery, sensors, and advanced materials, enabling innovations in various fields.
- **Without Plastic**: Progress in nanotechnology would be limited, as the lightweight and customizable nature of plastics is crucial for many applications. Alternatives may not offer the same functionality or versatility.

10. Space Exploration

- Plastics are used in spacecraft components, insulation materials, and lightweight structural elements, contributing to the overall efficiency and performance of space missions.
- **Without Plastic**: Spacecraft would be heavier and more challenging to design and launch, potentially increasing costs and hindering missions. The absence of lightweight materials could limit the scope of exploration.

11. Telecommunication

Apart from above, in the modern telecommunication industry, plastic plays a crucial role in various applications, including the innovative field of Li-Fi (Light Fidelity). Here's how plastic contributes:

- As Plastic cables are less prone to breaking and can withstand more bending, making them suitable for various environments, *the optical fibres (POFs)* are used for short-distance data transmission.

- Plastic enclosures protect sensitive electronic components in telecommunications devices from environmental factors such as moisture, dust, and physical damage.
- Plastic is often used to create connectors and interfaces for devices, allowing for easy assembly and disassembly. This is crucial for network maintenance and upgrades.
- The integration of plastic in the housing of mobile devices allows for better signal reception, which is essential for both Wi-Fi and emerging Li-Fi technologies.
- In Li-fi Technology, LEDs are often housed in plastic fixtures. The lightweight and mouldable nature of plastics allows for innovative designs that optimize light output.
- Many routers use plastic to reduce weight and cost, providing an efficient and effective housing for antennas and circuit boards.

5G technology

4G technology operated with more widely spaced cell towers that cover broader areas. In contrast, 5G technology is designed to support a massive number of devices—potentially up to 1 million devices per square kilometer—making it suitable for IoT (Internet of Things) applications. It utilizes a broader range of frequencies, including bands up to 100 GHz, which allows for faster data speeds and higher capacity. 5G promises higher speeds, with maximum speeds exceeding 10 Gbps.

To establish this technology, the number of required cell towers is significantly larger. Plastic components play a crucial role in this field. They are used in the construction of antenna enclosures, which are essential for 5G base stations.

As 5G relies on a dense network of small cells for coverage, plastic components enable compact and versatile designs that fit into urban environments without disrupting their appearance. Additionally, many 5G-enabled devices, such as smartphones and IoT sensors, incorporate plastic in their construction.

Li-fi Optical Fibre 5G Cell Tower

Chapter 6
Global Legislative Efforts on Reducing Plastic Pollution

Plastic pollution has emerged as one of the most pressing environmental challenges of our time, threatening ecosystems, wildlife, and human health.

Hence, governments around the world have recognized the urgent need for comprehensive legislative measures to mitigate this issue. From bans on single-use plastics to incentives for biodegradable alternatives, global legislative efforts to reduce plastic pollution reflect a commitment to sustainability and the protection of the environment.

United States of America

- **Plastic Bag Bans (2007–Present):** The USA saw the rise of anti-plastic bag campaigns starting in San Francisco, which was the first U.S. city to ban plastic bags in **2007**. This movement quickly spread across the country.

- **Microbeads Ban (2015):** The **Microbead-Free Waters Act** of 2015 marked a significant legislative action, banning the use of plastic microbeads in personal care products, such as soaps and exfoliants. This law helped raise awareness about the impact of microplastics on marine ecosystems and human health.

United Kingdom

- **Plastic Bag Charge (2015):** In **2015**, the UK introduced a 5p charge on single-use plastic bags, leading to an 85% reduction in their use by 2018. This initiative was a significant step in reducing plastic pollution across the UK, and it inspired similar policies in other countries.
- **Microbead Ban (2018):** Following in the footsteps of the USA, the UK banned microbeads in personal care products in 2018, demonstrating a growing awareness of microplastic pollution.

The ban was introduced under *The Environmental Protection (Microbeads) (England) Regulations 2017* and was extended to Scotland, Wales, and Northern Ireland.

European Union (EU)

- **Comprehensive Plastic Strategy (2018):** In 2018, the EU adopted a comprehensive **EU Plastics Strategy**, which included a goal to make all plastic packaging recyclable by 2030 and to reduce single-use plastics. The strategy addressed everything from plastic production to plastic waste management.
- **Single-Use Plastics Ban (2019):** In **2019**, the European Parliament passed a law banning single-use plastics, such as straws, cutlery, and plates, across all EU member states. The ban took effect in **2021**, making it one of the most extensive efforts globally to reduce plastic waste.
- In 2019, the **European Chemicals Agency (ECHA)** submitted a proposal to restrict the use of intentionally added microplastics (including microbeads) in products across the EU. This proposal was part of the **EU REACH Regulation**, which focuses on the Registration, Evaluation, Authorisation, and Restriction of Chemicals. The proposal aims to prevent the release of around 500,000 tonnes of microplastics into the environment over 20 years.
- The ECHA proposal specifically targets microplastics, including microbeads, that are intentionally added to products such as cosmetics, detergents, paints, and agricultural products.

- Several EU member states have introduced **national bans** on microbeads, particularly in cosmetics and personal care products. Countries such as **France**, **Sweden**, **Italy**, and **Belgium** have banned microbeads in rinse-off cosmetics like scrubs and toothpaste. These national regulations align with the broader EU goals of reducing plastic pollution.

Canada

- **Proposed Ban on Single-Use Plastics**: In 2021, the Canadian government announced its intention to ban harmful single-use plastics by the end of that year. This includes plastic bags, straws, and cutlery, aiming to reduce plastic waste in landfills and oceans.
- **Canadian Plastics Waste Reduction Strategy**: This strategy includes a commitment to ensure that all plastic packaging is recyclable or compostable by 2030. The strategy also emphasizes improving waste management and recycling infrastructure across the country.
- **Microbeads Regulations under the Canadian Environmental Protection Act, 1999 (CEPA):**

1. January 1, 2018: The manufacture and import of toiletries containing microbeads were prohibited.
2. July 1, 2018: The sale of toiletries containing microbeads was prohibited.

These regulations applied to personal care products containing microbeads, including Exfoliating Products (face wash, body wash, scrubs), Toothpaste, cleansers and Cosmetics.

Brazil

National Solid Waste Policy (2010)

- This law establishes guidelines for waste management, including the promotion of recycling and the reduction of waste generation. It encourages municipalities to implement selective waste collection and aims to reduce the amount of waste sent to landfills.

Law on Packaging (2020)

- This legislation establishes extended producer responsibility (EPR) for packaging waste, requiring manufacturers to ensure that their products are recyclable and that they manage the collection and recycling of packaging materials.

China

- **Yangtze River Pollution (2000s–Present):** China, the world's largest plastic producer, has also been a major source of plastic pollution. Studies show that the **Yangtze River** is responsible for dumping significant amounts of plastic into the ocean. In response, China implemented a **nationwide plastic bag ban** in **2008**, requiring retailers to charge for plastic bags. Despite the ban, challenges in enforcement have persisted.

- **Waste Import Ban (2018):** China was once the largest importer of the world's plastic waste, processing nearly half of global waste. In **2018**, China shocked the world by banning the import of plastic waste, forcing countries to reconsider their waste management strategies and creating a global plastic waste crisis.

Gulf Countries (UAE, Saudi Arabia, Qatar, etc.)

- **High Plastic Consumption Rates:** Gulf countries, such as **UAE, Saudi Arabia**, and **Qatar**, have some of the highest per capita rates of plastic consumption in the world. Disposable plastics are widespread in everyday life, largely due to oil-based economies that make plastic production inexpensive.

- **Plastic Bag Bans:** In **2022**, **UAE** implemented a ban on single-use plastic bags in **Abu Dhabi** and **Dubai**, signaling a shift in tackling plastic pollution. Other Gulf countries like **Saudi Arabia** have introduced similar regulations, including the banning of lightweight plastic bags in certain areas.

Australia

- **Marine Plastic Pollution:** Australia has a vast coastline, and plastic pollution has become a significant issue for its marine environment. The **Great Barrier Reef** and other coastal ecosystems are threatened by plastic debris. Research has shown that marine wildlife, such as sea turtles, is increasingly affected by plastic ingestion and entanglement.
- **Plastic Bag Bans (2018):** In **2018**, Australian states like **Queensland** and **Western Australia** banned single-use plastic bags, following the lead of **South Australia**, which had introduced a ban in **2009**. These initiatives were part of a broader national push to reduce plastic waste.
- **National Plastics Plan (2021):** Australia launched its **National Plastics Plan** in **2021**, which aims to phase out certain problematic plastics by 2025 and improve recycling rates across the country.

Japan

- **Plastic Bag Charge (2020):** In **2020**, Japan introduced a nationwide **charge for plastic bags** in retail stores, marking the first major policy aimed at reducing plastic waste. This move came in response to both domestic environmental concerns and international pressure, as Japan has been criticized for its high plastic use despite its advanced waste management systems.
- **Marine Plastic Commitments:** Japan, as a major maritime nation, has recognized the threat of plastic pollution to the oceans. In **2019**, during the **G20 Summit**, Japan pledged to reduce plastic waste and improve recycling efforts as part of its **Osaka Blue Ocean Vision**.

Rwanda

- **Plastic Bag Ban (2008)**: Rwanda implemented one of the world's strictest bans on plastic bags in 2008. This comprehensive policy prohibits the production, importation, and use of plastic bags, significantly reducing plastic waste in the country.
- **Awareness Campaigns**: The Rwandan government has conducted public awareness campaigns to educate citizens about the environmental impact of plastic waste and promote sustainable alternatives.
- **Environmental Clean-Up Initiatives**: Rwanda holds monthly community clean-up days known as "Umuganda," where citizens come together to clean their neighbourhoods, fostering a culture of environmental responsibility.
- **Promotion of Eco-Friendly Alternatives**: The government has encouraged the use of biodegradable and reusable products, supporting local businesses that produce sustainable alternatives to plastic.

New Zealand

- In 2019, New Zealand implemented a ban on single-use plastic shopping bags, aiming to reduce plastic pollution and promote the use of reusable bags. The ban applies to all retailers and includes a wide range of plastic bags.
- The **Waste Minimisation (Microbeads) Regulations 2017** came into effect on **June 7, 2018**.

Singapore

- **Plastic Bag Charge**:

In 2023, Singapore announced a national initiative to implement a charge on single-use plastic bags. This policy encourages consumers to bring their own reusable bags and aims to reduce the consumption of plastic bags.

Thailand

- **Plastic Bag Ban**:

In January 2020, Thailand began implementing a ban on single-use plastic bags in major retail stores and shopping centers. This initiative aims to reduce plastic pollution and encourage consumers to use reusable bags.

India

Swachh Bharat Mission (Clean India Mission)

- Launched in 2014, this initiative promotes cleanliness and waste management across the country, including efforts to reduce plastic waste through public awareness campaigns.

Plastic Waste Management Rules (2016):

- The **Plastic Waste Management Rules** were introduced in **2016** to promote waste management and recycling practices, which indirectly impacts the use of microbeads in products. These rules require manufacturers to adhere to environmentally friendly practices and promote the reduction of plastic usage.
- The rules were amended in **2018** and **2021**, expanding the scope of regulations on plastic waste and promoting sustainable alternatives.

Draft Notification on Microplastics (2020):

- In **2020**, the **Ministry of Environment, Forest and Climate Change (MoEFCC)** released a draft notification aimed at regulating microplastics, which includes microbeads, in various products. This notification is part of India's broader strategy to reduce plastic waste and its environmental impact.
- The draft notification proposed guidelines for the use of microplastics in cosmetics and personal care products, indicating a move towards regulating microbeads specifically.

Ban on Single-Use Plastics (2021)

- The Indian government announced a nationwide ban on single-use plastics, effective July 2022. This includes items like plastic bags, straws, and cutlery. States and local authorities are also encouraged to enforce this ban.

It is important to point out that not only the countries mentioned above, but also many others around the world, have implemented laws to reduce and prevent plastic usage in public spaces. However, it is not feasible to cover data from all countries; this overview highlights model implementations from a few nations aimed at raising awareness. This does not imply that the countries not listed have not taken steps to control plastic usage.

Therefore, it is clear that global legislative efforts to reduce plastic pollution are significant and vital in addressing one of the most pressing environmental challenges of our time. By enacting bans on single-use plastics, regulating production and waste management, supporting biodegradable alternatives, and fostering international collaboration, governments are making crucial strides toward a more sustainable future.

> *"You cannot get through a single day without having an impact on the world around you. What you do makes a difference, and you have to decide what kind of difference you want to make"* - Jane Goodall is a renowned British ethologist, primatologist, and anthropologist.

Chapter 7
The Shift to Biodegradable Materials Across Industries

As highlighted in the last chapter, many governments worldwide are enacting laws and regulations to combat plastic pollution. These policies often include bans on single-use plastics, incentives for using biodegradable alternatives, and stringent waste management regulations. Consequently, various industries are increasingly turning to biodegradable materials as sustainable alternatives.

Biodegradable materials are substances that can be broken down by natural processes, typically through the action of microorganisms, into non-toxic components that can re-enter the ecosystem without causing harm.

This shift towards biodegradable materials is driven by a combination of regulatory pressures, consumer demand for sustainability and the urgent need to reduce waste. The exploration of biodegradable materials across various industries reflects a growing recognition of the necessity for sustainable alternatives to traditional plastics.

Various industries contribute to plastic pollution and here is a breakdown of the major sectors responsible for this issue, along with the impact of increasing plastic pollution on the environment and the alternatives adopted by these industries

1. Packaging Industry

- **Contribution**: The largest contributor to plastic waste, accounting for about 40% of all plastic produced. This includes single-use items like bags, wrappers, and containers.
- **Impact**: Plastic packaging often ends up in landfills or the ocean, contributing significantly to environmental pollution.

Plastic packing materials

Alternatives:

- **Biodegradable Plastics**: Made from materials like corn starch, these break down more easily than traditional plastics.
- **Mushroom Packaging**: Created from agricultural waste and mycelium, it's compostable and biodegradable.
- **Seaweed-Based Packaging**: Flexible and edible packaging made from seaweed extracts.
- **Paper and Cardboard**: Used for boxes, bags, and wraps, often coated with plant-based materials for moisture resistance.

Mushroom packing

Seaweed packing

2. Textiles and Fashion Industry

Contribution: Synthetic fibres, such as polyester and nylon, contribute to microplastic pollution. The fashion industry is a major source of plastic fibres released during washing.

Impact: Washing synthetic clothing can release thousands of micro fibres per wash, which enter waterways and harm marine life.

Synthetic fibre clothes

Alternatives:

- **Organic Cotton**: Grown without synthetic pesticides or fertilizers.
- **Hemp**: A durable and fast-growing plant used for fabrics that require less water and chemicals.
- **Tencel (Lyocell)**: A biodegradable fibre made from sustainably sourced wood pulp.

Hemp cloth   Cotton cloth   Lyocell cloth

3. Consumer Goods

- **Contribution**: Household products, electronics, and personal care items often contain plastic components or packaging.
- **Impact**: Single-use items like razors, toothbrushes, and plastic bottles contribute to landfill waste and ocean pollution.

Plastic House hold items

Alternatives:

- **Compostable Tableware**: Made from materials like sugarcane (bagasse), bamboo, or palm leaves.
- **Reusable Containers**: Emphasizing the use of metal, glass, or silicone containers for takeout and storage.
- **Edible Cutlery**: Forks and spoons made from grains or other food materials that can be eaten after use.

Sugarcane Bagasse Tableware   Bamboo Tableware   Edible Cutlery

4. Construction and Building Materials

- **Contribution**: Plastics are used in various building materials, such as insulation, pipes, and flooring.
- **Impact**: Although not typically single-use, plastic waste from construction can lead to significant pollution if not properly managed.

Plastic flooring

PVC pipes

Alternatives:

- **Bamboo**: A fast-growing and strong material used for flooring, framing, and more.
- **Earth-Based Materials**: Materials like rammed earth, straw bales, and cob for sustainable building.

Bamboo flooring

Earth based flooring

5. Agriculture

- **Contribution**: Plastics are used in agricultural films, greenhouse covers, and packaging for fertilizers and pesticides.
- **Impact**: Agricultural plastic waste can lead to soil contamination and runoff into waterways.

Greenhouse

Fertiliser

Alternatives:

- **Compostable Mulch Films**: Made from biodegradable materials to cover crops, reducing plastic waste in fields.
- **Natural Fibre Twine**: Using jute or sisal twine instead of plastic twine for tying plants

Compostable mulch film   Jute twine

6. Healthcare

- **Contribution**: Medical plastics, including single-use syringes, gloves, and packaging, contribute to significant plastic waste.
- **Impact**: Improper disposal of medical plastics can lead to environmental hazards and increase pollution.

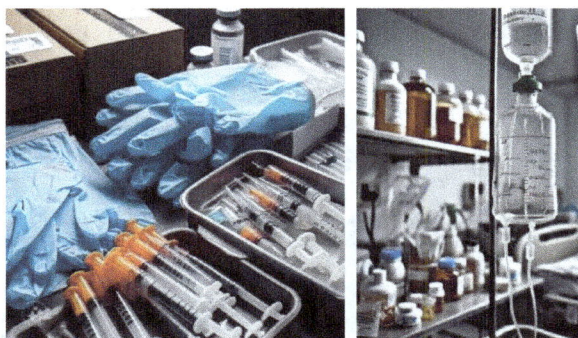

Disposable items

Alternatives

- **Biodegradable Materials:** Products made from materials like polylactic acid (PLA) that decompose more quickly than traditional plastics.

- **Glass:** Used for containers, vials, and syringes; glass is reusable and recyclable.

7. Fishing Industry

- **Contribution**: Lost or discarded fishing gear, known as "ghost gear," contributes to plastic pollution in oceans.
- **Impact**: Ghost gear can entangle marine wildlife and contribute to habitat destruction.

Alternatives

- **Biodegradable Fishing Gear:** Nets and lines made from materials like biodegradable polymers that decompose over time, reducing environmental impact.
- **Natural Fibres:** Use of materials such as cotton, jute, or hemp for nets and traps that are more environmentally friendly.

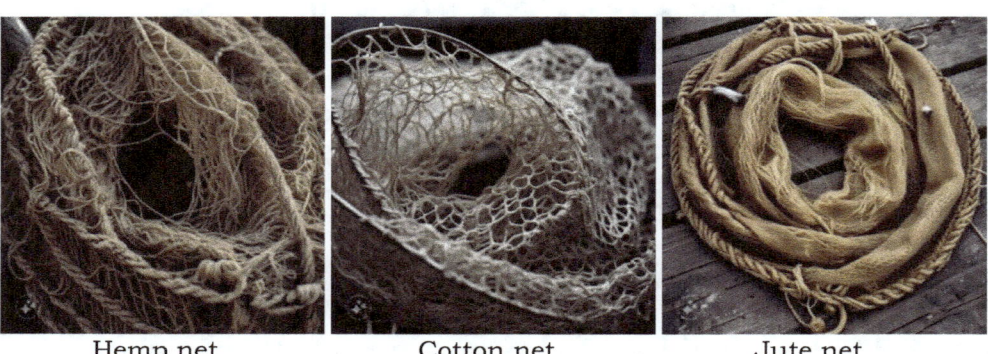

Hemp net   Cotton net   Jute net

The exploration of biodegradable materials across various industries reflects a growing recognition of the urgent need for sustainable alternatives to traditional plastics. This shift is not merely a trend but a necessary response to the escalating plastic waste crisis, which poses significant threats to our environment and public health.

The plastic waste crisis is a complex and multifaceted problem that requires sustained efforts from governments, industries, communities, and individuals. By implementing a combination of strategies that focus on prevention, reduction, recycling, and innovation, it is possible to make meaningful progress in tackling plastic pollution over time.

While significant progress has been made, challenges remain in terms of production costs, performance, and consumer acceptance. These obstacles highlight the need for continuous education and awareness initiatives to foster a deeper understanding of the benefits of biodegradable materials among consumers and businesses alike.

As industries continue to innovate and develop biodegradable solutions, collaboration among manufacturers, researchers, and policymakers will be essential to overcome these challenges and promote a more sustainable future.

Chapter 8
The Future – Challenges and Solutions

8.1 Earth Exists Because of Extinction

The Cycle of Life and Death in Nature and the Problem of Plastic

In nature, life and death are intricately linked, maintaining ecological balance. Every organism, from tiny leaves to massive trees and all animals, participates in this cycle. When they die, their bodies decompose, returning essential nutrients to the soil, enriching it and fostering new life. However, human-made materials like plastic disrupt this cycle and harm the environment.

Natural Decomposition and Its Importance

When trees shed leaves or animals die, decomposition begins. Microorganisms, such as bacteria and fungi, break down organic matter into simpler substances, creating humus—a rich soil component. This humus enhances soil fertility and supports plant growth. In this cycle, nothing goes to waste; the death of one organism fuels the life of another. Leaves, animal remains, and even human bodies ultimately nourish future generations.

Plastic: A Man-Made Product

In stark contrast, plastic is a synthetic material made from chemicals that does not decompose naturally. When discarded, plastic can linger in the environment for hundreds or thousands of years. Instead of enriching the soil, plastic waste accumulates in landfills, rivers, and oceans, creating significant ecological problems.

Plastic pollution harms wildlife, disrupts natural processes, and degrades soil and water quality. Unlike organic matter that provides nutrients, plastic breaks down into microplastics, which can contaminate the environment and enter food chains, posing risks to both wildlife and humans.

The Consequences of Plastic's Persistence

The persistence of plastic presents a formidable environmental challenge. While nature renews life continuously, plastic contributes non-degradable waste to ecosystems. For instance, a fallen leaf decomposes and enriches the soil, while a plastic bag can remain intact for centuries, blocking sunlight and hindering plant growth.

As plastic accumulates in landfills and habitats, it threatens the earth's ability to regenerate naturally, emphasizing the urgent need to address plastic pollution for a sustainable future.

8.2 Tips to Reduce Microplastic Exposure

The Microplastic Menace: A Simple Solution in Boiling Water

The Study's Findings

A team of researchers from Guangzhou Medical University and Jinan University in China conducted an experiment to test the efficacy of boiling water in removing microplastics. They added nanoplastics and microplastics to waters with varying levels of hardness and then boiled them for five minutes. The results were striking: boiling reduced the concentration of nanoplastics by 25% in soft water and a remarkable 90% in very hard water.

How Boiling Helps

The key to this process lies in the minerals present in hard water, particularly calcium carbonate. As the water temperature rises, these minerals form limescale, which traps the microplastics. By removing the limescale, individuals can significantly reduce their microplastic intake. To maximize the effectiveness of this method, the researchers recommend using non-plastic kettles and stainless steel filters.

Implications and Future Directions

The study's findings have important implications for public health. Microplastics have been linked to an increased risk of heart attacks, strokes, and even dementia. While the causal relationship between microplastics and these conditions is still being researched, reducing exposure to microplastics is a prudent step. The researchers believe that boiling water could be a viable long-term strategy for reducing global exposure to microplastics

A Traditional Practice with Modern Benefits

Interestingly, boiling water is an ancient tradition in some Asian countries, including China, Vietnam, and Indonesia. This practice not only purifies water but also reduces microplastic content. By adopting this simple habit, individuals can take a proactive approach to minimizing their exposure to these potentially harmful particles.

Conclusion

While further research is needed to fully understand the effects of microplastics on human health, boiling water offers a straightforward and effective way to reduce exposure. By incorporating this practice into daily life, individuals can take a significant step towards protecting their health and well-being. As the researchers conclude, "Drinking boiled water is a viable long-term strategy for reducing global exposure to microplastics.

In addition to methods like boiling and filtering water to reduce microplastics, you can further lower your microplastic intake by following these steps:

1. **Use Glass Containers**: Prepare baby formula and store food in glass containers instead of plastic ones to avoid microplastic contamination.

2. **Choose Less Processed Foods**: Avoid highly processed foods, such as chicken nuggets, as they may contain more microplastics due to contact with plastic equipment during production.

3. **Replace Plastic Cutting Boards:** Use wooden cutting boards instead of plastic ones. Cutting on plastic boards can release tiny plastic fragments into your food.

4. **Avoid Heating Plastic**: Never heat food in plastic containers, such as in microwaves, as heat can cause plastics to break down and release microplastics into your food.

5. **Switch Tea Bags**: Opt for paper tea bags or loose-leaf tea instead of plastic tea bags to prevent plastic particles from leaching into your drink.

6. Use Reusable Water Bottles: Replace single-use bottled water with reusable bottles made from stainless steel or glass.

7. Filter Your Water:

Install home water filters certified to reduce microplastics in your drinking water.

8. Wash Clothes Wisely:

Reduce the frequency of washing synthetic fabrics. Wash full loads of laundry and air-dry clothes to minimize the release of microfibers.

9. **Opt for Natural Fabrics**: Gradually replace synthetic fabrics such as polyester and nylon with natural alternatives like cotton, wool, or linen.

10. **Avoid Plastic-Based Furniture**: Choose furniture made from natural materials instead of plastic-based options like polypropylene fabric.

11. Checking Ingredient

Check ingredient lists for microplastics. Look for products labeled 'microplastic-free' or certified by environmental organizations.

Use smartphone apps that scan cosmetic ingredients and highlight harmful substances.

Many smartphone apps, such as **Yuka**, **Think Dirty**, **SkinSAFE**, and **INCI Beauty**, can scan and analyze these ingredient lists to help you understand their safety and effects.

Even though microbeads have been banned in many countries, we still need to be more conscious when buying these products in the market. The most common types of plastics used for manufacturing microbeads include:

- *Polyethylene (PE)*
- *Polypropylene (PP)*
- *Poly methyl methacrylate (PMMA)*
- *Nylon*
- *Polyethylene terephthalate (PET)*
- *Acrylates copolymer*
- *Polystyrene (PS)*

Conclusion: Moving Toward a Cleaner Future

The cosmetic industry is changing. With increasing regulations and consumer awareness, we are moving towards a future with fewer plastic-based ingredients. Make informed choices—because beauty should never come at the cost of our health or the environment.

By adopting these simple yet effective tips, individuals can significantly reduce their exposure to microplastics and contribute to a healthier environment.

8.3 Our Responsibilities

Nowadays, digital photography has replaced the need for plastic film, allowing for instant image capture *and* editing without the use of physical film. Likewise, CDs, DVDs, and plastic storage media have been largely replaced by digital downloads and streaming services, eliminating the need for physical discs. In the future, there may be a way to eliminate plastics in every field, paving the way to live safely and securely. However, as of now, the huge accumulation of plastic must be addressed.

Even though governments are implementing measures to reduce plastic usage, individuals also have a crucial role to play.

"We do not inherit the Earth from our ancestors; we borrow it from our children."

This means that our actions today will shape the world of tomorrow. Instead of viewing the Earth as a mere inheritance,

we should recognize it as a trust that we must nurture and protect. This perspective encourages sustainable practices, urging us to consider the legacy we leave for future generations.

Here are some responsibilities we can adopt to contribute to a more sustainable future:

1. Be Mindful of Single-Use Plastics

- **Avoid disposable plastic items** like straws, cutlery, and plastic bags. Opt for reusable alternatives such as metal straws, bamboo cutlery, and cloth bags.

2. Opt for Reusable Products

- Invest in **reusable containers**, bottles, and coffee cups to cut down on the need for single-use plastics.
- **Cloth bags** are a great alternative to plastic shopping bags. Carry one with you to reduce plastic bag usage at stores.
- Refill water bottles and use **reusable lunch boxes** instead of disposable ones.

For instance:

- Carrying household stainless steel tumblers and metal containers when you go on a picnic is a great way to reduce plastic use and minimize environmental impact. By avoiding disposable items, you're helping to keep nature clean and conserve resources.
- Making reusable carry bags from waste cloth is a fantastic way to repurpose fabric that would otherwise end up as waste. It's simple, eco-friendly, and a great way to promote sustainability. So, when you go shopping, take your cloth bag and avoid plastic carry bags.

3. Support Companies with Sustainable Practices

- Choose brands and companies that use sustainable packaging, such as biodegradable, compostable, or recyclable materials.
- Buy in bulk to reduce the amount of plastic packaging used in smaller quantities.

4. Reduce Plastic in Food Choices

- Minimize packaged food purchases; fresh fruits and vegetables typically come with little or no packaging.
- Support local, sustainable food sources to reduce the need for plastic-wrapped goods.

- When eating out, decline plastic takeout containers and request your food to be packed in paper or reusable containers.

5. Educate and Advocate

- Spread awareness within your community about the harmful effects of plastic and the importance of reducing consumption.
- Support policies and initiatives aimed at reducing plastic production, such as plastic bag bans or extended producer responsibility (EPR) programs.

6. Choose Eco-friendly Personal Care Products

- Use personal care items like bar soap, bamboo toothbrushes, and plastic-free shampoo bars.
- Be mindful of microplastics in personal care products (e.g., certain exfoliants) and opt for natural ingredients instead.

7. Be Conscious of Your Own Consumption

- Take a hard look at your daily routines and habits—identify areas where you can minimize plastic use.
- Track your plastic consumption to find where you can make the biggest impact, and set personal goals to reduce waste.

8. Building Awareness: Teaching Children About Plastic and Our Environment

Finally, today's children are the pillars of the future and will shape the world to come. Therefore, it is essential to educate them about the persistent issues related to plastic pollution.

One important concept is that everything in this world is interconnected. For example, trees provide oxygen and absorb carbon dioxide, while we breathe in oxygen and exhale carbon dioxide. In this way, trees and humans are deeply interconnected—trees can be seen as part of our lungs. Likewise, if the atmosphere, rain, soil, and sea are affected, they can become toxic, impacting not just the environment but all of us as well.

Educating children can be done in many ways, and I suggest a fun activity involving the counting of plastic items at home. This activity helps children visualize the extent of plastic use in their daily lives and think critically about alternatives.

Instructions for the Activity:

1. Ask children to go around the house and count how many plastic items they can find. Encourage them to look in all areas of the home: the kitchen, bathroom, living room, and bedrooms. They should not leave out anything, from small items like buttons in shirts to larger items like cars.
2. They should make a list of the plastic items they find.

After the children have completed their lists, ask them to reflect on their findings:

- How many plastic items did they find in their home?
- What type of plastic is each item made of?
- How long does each type of plastic take to decompose?
- What alternatives to plastic can they think of?

Next, engage the children in a discussion about what happens when plastic starts to decompose. Ask them to consider the effects of plastic on the environment, wildlife, and ecosystems during its long decomposition process.

Make them aware of microplastics and microbeads, and prepare them to use alternatives to plastics. It is our duty to hand over this world to today's children in the most responsible way.

8.4 Hopeful Advances in Biodegradable Plastics

So far, we have explored the harmful effects of microplastics that result from the breakdown of larger plastic items and discussed our responsibilities towards protecting the Earth. However, there is still hope. Ongoing research is leading to the development of new biodegradable alternatives that can replace traditional, non-biodegradable plastics. One such promising material is PHA (Polyhydroxyalkanoates).

PHA (Polyhydroxyalkanoates)

PHA is a type of biodegradable plastic produced through the bacterial fermentation of renewable biomass sources, such as sugarcane, potato starch, vegetable oils, or other organic materials.

Production Process:-

Bacteria are provided with raw materials like sugarcane, potato starch, or vegetable oils as their food source. The bacteria consume these materials and convert them into PHA, which is stored as granules inside their cells.

The PHA is extracted from the bacterial cells using solvents like chloroform or ethanol. It is then purified to remove impurities. The purified PHA is processed into pellets or films and can be moulded into various products using traditional plastic manufacturing techniques like injection moulding or extrusion.

Advantages

Biodegradable: PHA completely breaks down into water, carbon dioxide, and biomass in natural environments, making it environmentally friendly. Safe and Non-toxic: It is suitable for medical applications and food packaging.

As research continues, it is hopeful that even more advanced and sustainable biodegradable materials will emerge. In the future, items made from PHA and similar bioplastics may become commonplace, helping us move towards a cleaner, greener planet. Science is on our side, guiding us towards a future where biodegradable solutions can help us overcome the environmental challenges posed by traditional plastics.

"There is no Plan B because there is no Planet B." - Ban Ki-moon, Former U.N. Secretary-General

www.ingramcontent.com/pod-product-compliance
Lightning Source LLC
Chambersburg PA
CBHW062109220526

45471CB00010B/3661